Home Fires

How Things Worked

Robin Einhorn and Richard R. John, Series Editors

ALSO IN THE SERIES:
Ronald H. Bayor, *Encountering Ellis Island: How European Immigrants Entered America*

Home Fires

How Americans Kept Warm in the Nineteenth Century

SEAN PATRICK ADAMS

Johns Hopkins University Press | *Baltimore*

Printed in the United States of America on acid-free paper
9 8 7 6 5 4 3 2 1

Johns Hopkins University Press
2715 North Charles Street
Baltimore, Maryland 21218-4363
www.press.jhu.edu

Library of Congress Cataloging-in-Publication Data

Adams, Sean P.
Home fires : how Americans kept warm in the nineteenth century / Sean Patrick Adams.
pages cm. — (How things worked)
Includes bibliographical references and index.
ISBN-13: 978-1-4214-1356-3 (hardcover : acid-free paper)
ISBN-10: 1-4214-1356-6 (hardcover : acid-free paper)
ISBN-13: 978-1-4214-1357-0 (paperback : acid-free paper)
ISBN-10: 1-4214-1357-4 (paperback : acid-free paper)
ISBN-13: 978-1-4214-1358-7 (electronic)
ISBN-10: 1-4214-1358-2 (electronic)
1. Dwellings—Heating and ventilation—United States—History—19th century. 2. Heating—Social aspects—United States—History—19th century. 3. United States—Social conditions—19th century. 4. Social change—United States—History—19th century. 5. City and town life—United States—History—19th century. 6. Industrialization—Social aspects—United States—History—19th century. 7. United States—Social life and customs—19th century. 8. United States—Economic conditions—19th century. I. Title.
TH7216.U5A33 2014
697.0973'09034—dc23 2013037988

A catalog record for this book is available from the British Library.

Special discounts are available for bulk purchases of this book. For more information, please contact Special Sales at 410-516-6936 or specialsales@press.jhu.edu.

Johns Hopkins University Press uses environmentally friendly book materials, including recycled text paper that is composed of at least 30 percent post-consumer waste, whenever possible.

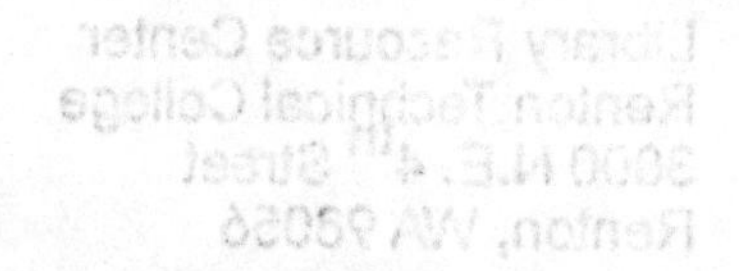

CONTENTS

PREFACE

When we flip the switch, turn the dials, or set the digital thermostat to keep our present-day homes warm, do we ever pause to think about the various people and institutions behind that action? Do we realize how that simple act is made possible by a host of economic, technological, and political networks busy making sure that ample energy is available for our comfort whenever we need it? Although Americans rarely contemplate these things—it's much easier just to flip the switch—there is a long and fascinating history that made home heating easy and inexpensive by the twentieth and twenty-first centuries. As part of the How Things Worked series, this volume explains why that story of heating is important, how it is deeply intertwined with the broader scope of American industrialization, and how the new industrial economy helped subdue one of the nastiest elements of life in the northern half of the United States: the winter cold.

But this transformation occurred neither instantaneously nor without struggle and conflict. I hope that readers of *Home Fires* will come to understand how the process of industrialization created an overlapping series of networks that grew over time and that the simple process of staying warm in the winter drew upon the efforts of thousands of Americans. Over the course of the nineteenth century, as this book explains, the long-held tradition of using wood fuel in fireplaces was replaced by coal-burning stoves and grates. Cheap heat came with immediate costs and benefits as well as an abiding dependence upon mineral fuel. Examining the changes in home heating offers us a way to explore exactly what "industrialization" meant for the average American family living in one of the many growing cities of the American North. It changed everyday life for those households, in some good ways and in some bad ways, as a diverse cast of economic actors—scientists, engineers, miners, philanthropists, merchants, day laborers, coal dealers, railroad executives, consumers both rich and poor—worked to keep the home fires burning. The changes in home heating, which I describe in the pages that follow as the rise of the "industrial hearth," wrote the first chapter in a long tradition of American dependence upon cheap and abundant fossil fuels.

Today we hear a great deal about changing that dependence. The overwhelming scientific evidence suggests that burning coal and oil has put all of humanity on a dangerous trajectory in terms of pollution, climate change, and the increasing scarcity of energy resources. We all want change for the better, but we must reckon with the massive economic, social, and political investments that American society has made in fossil fuel energy over the last two centuries. What price are Americans willing to pay to sustain their comfortable homes now and to secure a better energy future? History can't necessarily provide a definitive answer to this question, but the study of past experiences provides a better-informed perspective on the present and even the future. In other words, we need to know how this love affair with cheap mineral energy began and has been maintained before we can start thinking about how to improve the relationship. If we understand the contingent factors involved in the first mineral fuel revolution—the transition that serves as the central theme of this volume—we are better equipped to anticipate and enact the next major change in energy use.

You don't need to publish in a series called How Things Worked to know that the question of "how a book works" cannot be answered easily. Much like the coal that traveled from mine to the hearth in the nineteenth century, the twenty-first-century manuscript moves to and from various individuals and institutions, all of whom do their best to deliver the best end result. It is a complicated endeavor, to say the least, and one that requires a great deal of effort from a wide range of people. There are a number of them that I would like to thank for helping to make *Home Fires* possible; in fact this book would never have come to anything without their dedication and professionalism.

I'd first like to thank the series editors, Richard John and Robin Einhorn, who were encouraging at every step of the process of writing this book, from prospectus to final submission. Bob Brugger and his staff at Johns Hopkins University Press showed skill and diligence in working with me from conceptualization through completion. The University of Florida's Department of History provided an intellectual and professional home for me while I wrote this book, and I'd particularly like to thank my colleagues Jeff Adler, Jon Sensbach, Mitch Hart, Nina Caputo, Elizabeth Dale, Luise White, Bill

Link, Matt Gallman, Ida Altman, and Jeff Needell for various discussions and debates in the hallways, seminar rooms, and local restaurants that helped me sharpen my ideas. This book took shape at a particularly difficult time for the UF community, and the History Department continues to persevere as a nurturing place for scholars and teachers. I am proud to be a part of it, even though the mild Gainesville winters offer little opportunity for me to test my newly acquired expertise in home heating.

A wider community of historians also offered the kind of constructive criticism and encouragement that authors need to succeed. At various points of this project, I received help from a host of scholars. All of them were patient in hearing more about nineteenth-century home heating than they probably ever wanted to know. I'd like to thank Christopher Jones, Richard John, and the anonymous reader for Johns Hopkins University Press for their insightful read of the entire manuscript. Howell Harris, whose work on nineteenth-century stoves promises to be the definitive work on the subject, provided a generous reading of my own chapter on the subject; if I needed to know anything about the manufacturing and marketing of stoves, I turned to Howell. Thanks also to Andrew Arnold, John Brooke, Spencer Downing, Laura Edwards, Dan Feller, Stephen Mihm, Sharon Murphy, Larry Peskin, Jonathan Rees, Seth Rockman, Lee Vinsel, Conrad Edick Wright, and Wendy Woloson. This list is surely incomplete; many others helped me obtain research material, pushed me to move the argument in different directions, or provided other invaluable help. I deeply appreciate everyone's assistance.

Along with the bigheartedness of fellow scholars, I need to recognize some of the institutions that offered financial support for the research and writing on this project. Thanks very much to the Program in Early American Economy and Society at the Library Company of Philadelphia, the National Endowment for the Humanities, the Chemical Heritage Foundation, the Gilder Lehrman Institute, the Hagley Museum and Library, the Massachusetts Historical Society, and the University of Florida's College of Liberal Arts and Sciences. Portions of chapter 2 appeared in "Warming the Poor and Building Consumers: Fuel Philanthropy in the Early Republic's Urban North," *Journal of American History* 95 (June 2008): 69–94, and portions of chapter 3 appeared in "Soulless Monsters and Iron Horses: The Civil War, Institutional Change, and American Capitalism," in *Capitalism Takes Command: The Transformation of Nineteenth-Century America*, ed. Gary Kornblith and Michael Zakim (Chicago: University of Chicago Press, 2012), 249–276.

I'd like to dedicate this book to Juliana Barr, the most accomplished American historian in our family. She writes about completely different regions, different peoples, and with a different perspective than my own. Nonetheless, writing this book on home heating would have been impossible without her company. Juliana's presence warms my heart and soul completely, and she doesn't need an ounce of coal to do so.

Home Fires

Prologue

"WE HEAR OF Persons frozen to Death on both Rivers contending with the Ice and of multitudes who suffer in various Parts of the Town," New Yorker William Smith wrote in 1780. "No Wood can come from the other Side of the Water and tis said this Island will be totally disforested in a Week." As Smith and other residents of the occupied cities of the North hunkered down for long, cold, and trying winters during the American Revolution, they faced a crisis of growing proportions. The increasing scarcity of firewood in American cities had been evident during the colonial period, but the political and military developments exacerbated its impact. During the winter of 1775–76 Bostonians resorted to tearing apart John Winthrop's old house for firewood. When no wood was to be had, they burned horse dung to keep warm. The British army, which controlled the city, offered little assistance; one resident recalled that while the cold afflicted many Bostonians, "such was the inhumanity of our masters that [we] were even denied the privilege of buying the surplusage of the soldiers' rations." When the hard winter of 1779–80 hit, it was the coldest season in recorded history for the American colonies. Philadelphians roasted an ox on the iced over Delaware, and the British army crossed the frozen waterways of the Northeast without concern. No group felt the combined effects of fuel shortages and severe weather more keenly than the laboring classes and dependent poor, for whom the fuel shortages pinched already scarce household budgets and threatened to lead to sickness and death. "God have mercy on the Poor," William Smith continued in his 1780 diary. "Many reputable People lay abed these Days for Want of Fuel."[1]

Independence may have warmed the hearts of patriots, but it offered little relief from the cold. Growing urban populations, declining supplies of firewood, and seasonal unemployment during the winter months produced intense suffering in American cities, even after the British soldiers evacuated them. Among urban residents, the poor felt the effects of the prolonged fuel crisis the most. The price of firewood nearly tripled from 1754 to 1800, when measured against the cost of other necessities such as food, clothing, and housing. High prices led to a failure in the fight against cold weather, which made living in poverty in urban areas more dangerous to life and limb. Severe winters exacerbated persistent health problems facing the poor, as cold weather combined with fuel shortages made sickness even more deadly. Mortality statistics for what we now call hypothermia do not exist in the extant records, but undoubtedly the cold conditions amplified more commonly labeled disorders such as "decay" and "disability" in cities like Philadelphia. Although they suffered less immediate danger, affluent residents of American cities also complained of the increasing scarcity and high prices of firewood. Harvard officials bemoaned the "great scarcity of wood in the College" and offered leave to students who were "destitute of fewel." For all residents, the challenge of staying warm in the burgeoning cities of the United States raised a number of thorny issues for the new nation. What could be done to ameliorate the severe effects of winter among the urban populace? Who could provide remedies for both the immediate problems of the freezing poor and the more widespread impact of spikes in fuel prices? Why did firewood shortages occur more and more frequently? What should be done about this crisis in home heating?[2]

Although there were some tweaks to the physical structure of fireplaces, chimneys, and other home heating implements, the basic technology used to heat domestic spaces had changed very little in the five centuries before the American Revolution, and the major fuel, firewood, had been the preferred heating fuel for millennia. The problem early American cities faced had deep roots in the traditional construction of the hearth and its centrality to home heating. As their ancestors had done for thousands of years, Americans at the time of the Revolution burned wood to keep warm. The wood-burning hearth had a long history as the main heating source for households in the Western tradition. The notion of a hearth originally meant the practice of keeping an open fire in the center of a room that threw smoke and soot—for the most part—out of a hole in the ceiling. Eventually, the construction of large stone buildings like churches, monasteries, and castles in medieval Europe set the hearth into the walls of individual chambers and provided a flue for the escape of smoke. Incremental innovations in fireplace and chimney design provided less messy ways to burn wood, while retaining the large roaring fire that had become so closely associated with the idea of the hearth as the central focus of the household. By the time Europeans began colonizing North America, they brought with them two basic notions about heating a home. First, the hearth served as its symbolic center. Members of the household gathered around a fire in order to stay warm, to be sure, but such close proximity offered the opportunity to socialize and form lasting bonds over long winter nights. Second, American colonists brought along with them the tradition that a fireplace made of stone or brick was the proper location for the hearth. Residents of both large and modest homes used fireplaces to cook their meals, warm their bodies, and serve as the social center of the household. The idea of the hearth and the use of open fireplaces thus went hand in hand.[3]

The fuel crisis that rocked the cities of the American Revolution suggests that this time-honored tradition was evolving into a bad habit; scarcities of firewood plagued urban residents beyond the years of occupation mostly because the fireplace is an incredibly inefficient method of heating a room. When you burn wood in a fireplace, the irritating soot and smoke fly up the chimney, but so does the vast majority of heat. A great abundance of fuel, therefore, is needed to produce even a small increase in temperature. Even fully stoked, fireplaces were uneven, as they produced a withering heat in close proximity, while the radiant heat produced did very little to warm an entire room. Open fireplaces did not work well for home heating in the American climate even when plenty of firewood was to be found. The New York physician Cadwallader Colden noticed this problem as early as the 1720s. "I found Madeira wine (which is a very strong wine) frozen in the morning," he wrote on one particularly frosty day, "in a room where there had been a good fire all day till eleven at night." As much as Colden might find his frozen Madeira an unpleasant side effect of the cold, less affluent residents of larger colonial cities like Philadelphia, New York, and Boston found the winter's chill a much more serious challenge to their health and well-being. And as surrounding stocks of firewood gradually disappeared, this problem would only become more severe. Even in cities like Baltimore, in which the climate was somewhat more temperate, poor families suffered miserably in the cold season as firewood ran in short supply. This endemic crisis threatened the very core of urban life in the Early American Republic.[4]

If the root cause of this fuel crisis in home heating was the old-fashioned fireplace and chimney system, then perhaps some inventive minds could be put to work on the problem. After all, this was the period of the Enlightenment, when many time-honored assumptions concerning politics and society fell victim to innovations based on scientific principles, experimentation, and rational thought. Americans had carved a new form of political governance in the face of the traditional monarchy, so why not apply this innovative spirit to the old-fashioned hearth? In fact, two very bright American minds tackled the problem of fuel scarcity during the turbulent years of the American Revolution. They both happened to be named Benjamin, both spent their childhood enduring frosty Massachusetts winters, and both became influential thinkers and writers of the Enlightenment. Could their formidable intellects solve the heating fuel crisis faced by American cities?

The first inventor to work on this problem, Benjamin Franklin, hardly needs introduction. As a noted printer, scientist, philosopher, and politician, Franklin dominated the American intellectual landscape by the late eighteenth century. But one of his first major experiments tackled America's home heating problem. "Wood, our common Fewel, which within these 100 Years might be had at every Man's Door," Franklin wrote in 1744, "must now be fetch'd near 100 Miles to some towns, and makes a very considerable Article in the Expence of Families." His solution was published in a celebrated pamphlet describing the "Pennsylvanian Fire-Place," a cast-iron box that used an inverted siphon to draw the heat created by combustion through a series of baffles before making its way up through the chimney (fig. P.1). Although it is often referred to as the "Franklin stove," the original design was in fact a modification of existing fireplaces that sought to solve the problem of unevenly heating a room; as Franklin described the problem, "In severe Weather, a Man is scorch'd before, while he's froze behind." Since he conceived of heat as a kind of fluid, the basic notion behind the Pennsylvanian fireplace was to allow it to flow over the baffles and thus keep it in the room for a longer period of time. Franklin's inverted siphon drew the smoke up through the chimney, away from the iron plates, and thus the "air that enters the Room thro' the Air-box is fresh, tho' warm." Franklin argued that, once

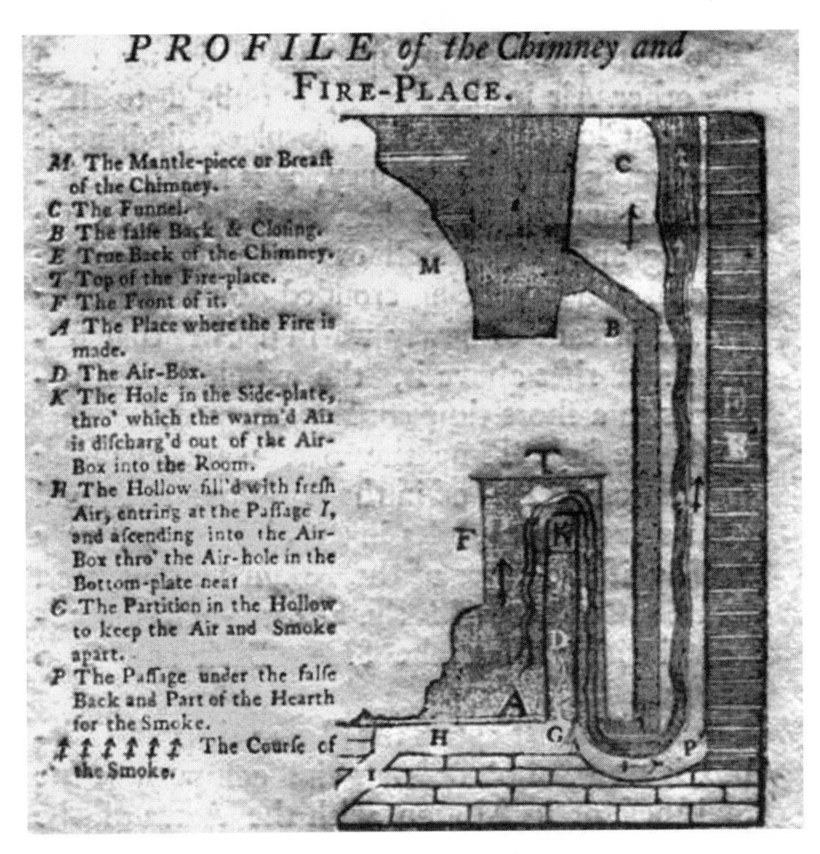

Figure P.1. The "Pennsylvanian fireplace," or "Franklin stove." This cross-sectional diagram from Franklin's 1744 pamphlet demonstrates how the Pennsylvanian fireplace worked in theory. In reality, though, the smoke and soot did not act precisely as Franklin envisioned. More practical designs of stoves, based less on theories of heated air and more on the practical experience of stove makers, ultimately triumphed in the American marketplace. Benjamin Franklin, *An Account of the New Pennsylvanian Fireplace* (Philadelphia: B. Franklin, 1744), 2

•

integrated into existing fireplaces, his device would reshape the hearth so that "People need not croud so close round the fire" and would provide a fuel savings such that "much less Wood will serve you, which is a considerable Advantage where wood is dear."[5]

Franklin's intent was to apply scientific methodology to the problem of home heating. "I suppose our Ancestors never thought of warming Rooms to sit in," he mused. "All they propos'd was to have a Place to make a fire in, by which they might warm themselves when acold." The Pennsylvanian fireplace was not wholly his own invention; Franklin acknowledged that he learned much about the characteristics of heat from a French book written in 1715 by Nicolas Gauger entitled *The Mechanics of Fire*. Inspired by Gauger's designs, Franklin wanted to improve the efficiency of the American hearth and, at the same time, test some theories he had about the movement of heated air. The problem was that it didn't heat rooms very well. The principle seemed perfect on paper; in real life the Pennsylvanian fireplace blew smoke back into the room and required constant attention. Franklin continued to tinker with stove designs throughout his life but could never solve the problem to his satisfaction. When he drew up plans for a freestanding cylindrical stove in the early 1770s, it also failed to make an impact in home heating markets. Yet Franklin's intellectual celebrity had increased so much by that time that Americans erroneously refer to most cylindrical stoves, particularly those that bulge in a potbellied fashion, as "Franklin stoves." Benjamin Franklin was a towering presence during the American Revolution and perhaps the first truly great American thinker, but historical accuracy requires that we differentiate between his invention and the innovations that came later. The Pennsylvanian fireplace failed to solve the problems of Early American home heating, and many "Franklin stoves" on the market today have little to do with Franklin's design.[6]

Another famous American named Benjamin also sought to confront the problem of open fireplaces. Benjamin Thompson was born in Massachusetts but proclaimed his loyalty to King George III during the American Revolution. Thompson served the Loyalist cause during the conflict and worked on improving gunpowder technology. This eventually earned him a knighthood and a place in London's scientific community. By the late 1780s, Thompson had relocated to Munich, where he worked on the reorganization of the Bavarian army under the patronage of Carl Theodore, the ruler of Bavaria at the time. As an elector of the Holy Roman Empire, Carl Theodore had the power to bestow titles of nobility, and Thompson's successes earned him an aristocratic title. In a swipe at what he believed was the short-sighted republican culture that ruled in the land of his birth, Benjamin Thompson named his title "Rumford," which was the former name of his old hometown, Concord, New Hampshire. Thus equipped with this name and title, the newly minted Count Rumford turned his attention to home heating in a series of essays that trumpeted his innovations in Munich. Count Rumford's technological writings specifically wedded the idea of heat efficiency to philanthropy; among his responsibilities was the improvement of Munich's poorhouses. "It is not to be believed what the waste of fuel really is, in the various processes in which it is employed in the economy of human life," Rumford wrote, "and in no case is this waste greater than in the domestic management of the Poor."[7]

Count Rumford focused on the "plague of a smoking Chimney" and the "great defects in open Fire-places" that he witnessed in his American and European travels. Like Franklin, he conceived of heat as a kind of liquid substance, which conventional open fireplaces allowed to escape rooms rather quickly. Consider heat, he argued in "Of Chimney Fire-Places, with Proposals for Improving Them to Save Fuel," as moving through the air like oil on water. If that substance can be delayed in its flight up the chimney, he concluded, then a fireplace might gain efficiency. The solution, he found, was to narrow the flue of the chimney and angle the rear of the fireplace, thus allowing the invisible "radiant heat" to be reflected into the room and the "combined heat," which included the smoke, soot, and harmful elements of fire, to flow out of the chimney. Rumford's calculations for just how to angle the "radiant heat" were quite precise; over the 1790s he published several essays on the subject with very specific construction directions and recounted his various experiments with heating rooms in the barracks and poorhouses of Munich. Like Franklin, Count Rumford sought to apply his intellectual curiosity to useful innovation, and the Rumford fireplace, like the Franklin stove, soon garnered scientific acclaim in both England and the United States. Most important, though, Count Rumford's ideas became associated with the philanthropic benefits of home heating innovation. Since the poor are naturally "ignorant, negligent, and wasteful," one American reviewer of Rumford's work noted in 1798, "fire which they use to warm themselves, serves commonly to crease the cold. Slight and cheap improvements in the structure of their chimneys will obviate this inconvenience."[8]

The Pennsylvanian fireplace, or the real "Franklin stove," and the Rumford fireplace are both examples of the values of the Enlightenment put into practice in the late eighteenth century in that they both attempted to use the knowledge acquired through experimental science toward practical ends that would improve the quality and comfort of human life. Unfortunately, the actual contributions of Benjamin Franklin and Count Rumford to improving American methods of home heating were quite limited. The Franklin stove and the Rumford fireplace were not innovations that reshaped everyday life for most Americans, as changing long-standing assumptions about the ways in which they could, or should, heat their homes required more than a few well-publicized designs. The Franklin stove never quite caught on in the American marketplace—perhaps because though it satisfied Franklin's intellectual curiosity about the way heated air theoretically traveled, it didn't heat rooms very well, and it was expensive. Some American homes adopted Rumford fireplaces by the 1790s, but even as the improvement saved on wood consumption in individual households, a smattering of narrower chimney flues hardly made a dent in the growing fuel crisis facing American cities. Instead of merely adopting a more efficient design, Americans needed to completely overhaul the widespread use of large, open fireplaces burning wood. They needed an entirely new *system* of heating their households, not a few inventions or improvements to existing practices.

Highlighting only the bold innovations in design is an easy but inaccurate way to sum up approaches to the problem of home heating. The broad interpretation of what exactly constitutes a Franklin stove is a great popular example of that tendency. Contemporary sources suggest that few of Franklin's ideas made their way into American marketplaces and homes, yet as the historical legend of Franklin grew over the decades, so did the idea that freestanding stoves must have been a successful invention of his making. Current biographers certainly reinforce this notion, even while they recognize his initial failures. One recent biography praises Franklin's "bulbous metal heater that survives to this day." Another still gives the old philosopher credit for streamlining his stoves by asserting that "what is today commonly known as the Franklin Stove is a far simpler contraption than what he originally envisioned." The most outright admission of the shortcomings of his claim that his stoves recirculated warm air efficiently throughout the house still praises Franklin in a roundabout way. "He tried various ways of achieving that effect, not all of which worked," a 2002 biography argued, "but neither do all of those that are still being tried." The heroic status of the stove inventor is a red herring because it leads us to the conclusion that great thinkers like Franklin or Rumford could offer a simple solution to the home-heating crisis in urban America. The American stoves that did solve the nation's first home heating crisis were much simpler than the Franklin or Rumford inventions, but their wholesale adoption was much more complicated than either of these two great thinkers might have imagined.[9]

Methods of home heating improved dramatically over the course of the nineteenth century, and American cities eventually conquered the problem of fuel scarcity that so bedeviled the Revolutionary generation. But in order to do so, they needed much more than the Enlightenment-inspired ideas of Franklin or Rumford; they needed an industrial economy with a dense network of market interactions. Ben Franklin's turn as patriot turned inventor and Count Rumford's role as the Loyalist who ended up caring about the hearths of the poor are good stories, but the real transformation of the American hearth is not a matter of explaining how society eventually caught up to their brilliant ideas but instead that of millions of small, almost imperceptible, everyday events; collectively these slowly but surely solved the nation's first energy crisis and paved the way for cheap and efficient home heating. Newfangled designs and theories about heating homes abounded in both scientific circles and the marketplace, to be sure, but this critical energy transition only took hold after a growing wave of everyday market decisions regarding fairly simple, boxlike stoves. By the late nineteenth century, heating an urban household was both cheaper and more efficient than it had been in Rumford or Franklin's day, and the story of how that came about must take into account decisions by thousands of consumers to purchase stoves and thousands of men and women learning to build and maintain a fire fueled by coal instead of wood. It needs to recognize the millions of dollars in capital and millions of hours of labor expended in building a massive fuel distribution network; the dirty and dangerous work done in America's blast furnaces, iron foundries, and coal mines; and the bare-knuckle negotiations between colliers, railroads, wholesalers, coal dealers, and customers. The overlapping industrial networks that came into being to facilitate these changes in home heating were the creation of all these agents, sometimes in concert, often in conflict with each other. This is a story of the rise of industrial capitalism, the immense complexity of its development over the course of the nineteenth century, and the ways in which those dense and turbulent systems were necessary to the creation of an industrial hearth in the urban North. By the advent of the twentieth century, the transformation of the hearth had helped usher in one of the most dynamic industrial societies in the world.

By linking one of the most traditional anchors of those homes—the hearth—with quite sophisticated and complex networks like those involved in the manufacture of stoves, coal mining and distribution, and the advent of new technologies like steam heat, this book provides insight into how industrialization was experienced by average Americans. The simple yet elusive quest to stay warm during the cold winters of the American North took more than a hundred years to be completed, albeit imperfectly, and it did not occur without major struggles, false starts, imperfect designs, and competing interests that accompanied the rise of industrial capitalism in the United States. Therefore, instead of offering a narrative account of invention—as great and colorful a story as Ben Franklin and Count Rumford provide—it's more fruitful to examine the process of change in home heating over the course of the nineteenth century in order to see what it reveals about the broader concept of "industrialization." What does this process actually entail? What do we mean by it? One recent study of the process in Great Britain suggests that industrialization's major distinction from previous models of economic change was placing technology at the forefront of transformations of everyday life. In providing a working definition of *technology* here, it is important to keep in mind that the concept embraces both the material improvements in the process of doing things and the social, political, and cultural systems necessary to enact those modifications (no technological change occurs of its own volition); both the physical and contextual aspects of technology come into play. It's not really the minds that invent but the minds that are persuaded to buy and use that are of vital importance. What the nineteenth-century evolution of home heating did with an industrial technology hinged on the decisions, workings, and ideology of the household. There were physical changes, too, but this particular phase of industrialization required the presence of material improvements and the emergence of wider geographical networks and technological systems to enable those changes to occur over a large population. In a sense, then, the transformation of the urban hearth offers us an understanding of how that kind of abstract transformation of the economy became quite tangible in nineteenth-century homes.[10]

For the most part, historians depict industrialization in the United States as a transformation of the workplace. This makes sense, as many of the revolutionary innovations that we have come to associate with an industrial economy occurred there, and its broad contours offer a now familiar narrative. By taking advantage of the division of labor, managers reorganized the shop floor in many trades and wrung new levels of productivity from their workers. Most significant for the story of home heating, the replacement of "organic" forms of energy like human, animal, or water power with the mineral energy that fueled steam or electrical power allowed an unprecedented increase in manufacturing capacity. The appearance of more sophisticated machinery furthered both of these trends, even as it undermined the skills and autonomy of American workers. Cities swelled in size as factories rose to take advantage of these trends, thus creating huge domestic markets that spurred demand for more factories. The responses to industrialization were myriad. Workers struggled to maintain power on the shop floor and resisted the replacement of their know-how and skill through various means, sometimes resorting to walkouts and strikes. Families negotiated a new wage economy with a variety of strategies, drawing women and children into the paid workforce. Consumers relished paying less for those goods created by increased productivity and lower labor costs, even as some bemoaned the standardization and lack of craftsmanship that accompanied the mass production of goods. In the end, industrialization transformed many long-standing assumptions about the way things worked: it changed forever the way labor was valued and rewarded, increased the material standard of living for most Americans, and completely reordered the economic landscape of the United States.[11]

This book seeks to reconstruct how industrialization profoundly changed an important aspect of everyday life, and the chapters that follow offer a focused but critical insight into this weighty transformation of the American home. But rather than starting on the factory floor and then chronicling how Americans responded to changes that occurred there, I explore the needs of households and the ways production methods responded to those demands. This perspective allows us to trace how entrepreneurs and innovators in home heating attempted—not always successfully—to make their imprint in this new marketplace. The links between the hearth and the ongoing process of industrialization are the focus here; such a perspective offers not an exhaustive account of the material changes in home heating but an interpretation of how it developed. In taking this view, we'll see the many achievements of the industrial economy: the ways in which technological innovations improved everyday life, the creation of new and improved notions of physical comfort, the development of an unprecedented range of choices for consumers. But the rise of the industrial hearth also reveals the shortcomings of that same process: the displacement of traditional workers, the environmental and economic changes within households wrought by burning coal, the persistent division of consumers into those who can afford new technologies and those who cannot, and power struggles within home heating networks. In the end, staying warm in nineteenth-century American cities involved a multitude of actors, institutions, and markets enmeshed in a series of overlapping networks that, when considered together, offer a unique look into how industrial capitalism took shape and altered everyday life during a critical period of its development.

In this book, I chronicle how changes in household heating unfolded in diverse locations like the boardroom, parlor, and tenement to reconstruct the ways in which technical knowledge spread among the urban populace, the methods by which entrepreneurs attempted to convince consumers that their product designs and fuel choices were superior, and the response to these overtures among wealthy, middle-class, and poor Americans. The volume re-creates the ways scientific and technological breakthroughs negotiated their way through institutions, entrepreneurs, and consumers and explains the struggles of urban families as they sought to adapt to shifts in heating technology in the ever-changing nineteenth-century industrial landscape. Using the challenge of staying warm in the industrializing North as a window into the complex world of energy transitions, economic change, and emerging consumerism, I use the formation of the new methods of home heating to trace links between structural transformations in the American economy and the experience of average Americans in negotiating this industrial makeover. This perspective gives us a unique view of the emergence of an industrial society from the ground up or, in this case, the hearth up. But the involvement of all of these different actors and institutions leads us to begin with one simple question from which we can start.

How did it work?

1 How the Industrial Economy Made the Stove

IN COLONIAL BOSTON, a massive elm stood at the corner of present-day Essex and Washington Streets. Residents called it the "Great Tree," and although it already served as an important landmark for the city, on August 14, 1765, Bostonians would remake it into a symbol of colonial resistance and American unity. As a protest against Parliament's passage of the Stamp Act, Bostonians hanged an effigy of a British official on one branch and an image of Satan holding a copy of the legislation on the other. Dissatisfaction with colonial policy reached a breaking point that day, and a crowd tore down the effigy and created a mock funeral procession that ended with the destruction of the Stamp Office. A month later, Bostonians placed a copper plate with the words "The Tree of Liberty" on the trunk of that same elm tree, and a symbol was born. All along the American coast, from Rhode Island to New York and as far as south as South Carolina, colonists gathered at their own designated Liberty Trees. Boston's served as a rallying point even after the repeal of the Stamp Act, as residents there celebrated the Liberty Tree's birthday with a holiday designed to foster unity within the community. Soon afterward, the idea of the Liberty Tree evolved into the Liberty Pole, an iconic symbol of republican government that persisted in American politics well into the nineteenth century.[1]

Although its star shone brightly for a while, things did not end well for Boston's original Liberty Tree. Following the outbreak of the American Revolution, outraged Loyalists chopped down the Liberty Tree in August of 1775. "After a long spell of laughing and grinning, sweating, swearing, and foaming," the *New England Chronicle* reported, "with Malice diabolical they cut down a tree *because* it bore the *Name of Liberty*." Soon only a stump remained. The Liberty Tree did exact some small revenge in death, as legend has it that one of its attackers died from a falling branch. The Liberty Tree, like many trees before and after it, was destined to become firewood. Reports that it provided fourteen cords are likely an exaggeration—a single tree with a thirty-inch trunk that stands ninety feet tall produces a little less than four cords of firewood—yet the idea that this adored symbol would end up in the fireplace of Boston's British commander, General William Howe, sent a clear message to Boston's revolutionaries. A Tory newspaper published a mocking soliloquy for the Liberty Tree that mused:

My trunk may be to fuel turn'd
By HOWE, be honor'd to be burn'd
That to him may warmth impart
Who oft himself's warm'd many a heart.
If ever there should be a shoot,
Spring from my venerable root,
Prevent, oh heaven! it ne'er may see,
Such savage times of liberty.[2]

The Liberty Tree was much more valuable as a political symbol of the American Revolution than it ever was in heating homes. Even if it had provided its legendary fourteen cords, this would have made only a small dent in the fuel shortage that plagued American cities before, during, and immediately after British occupation. After all, had Bostonians secured a piece of the Liberty Tree for their fireplaces, the majority of the heat would have escaped through the chimney. British officers and soldiers would have been familiar with burning bituminous coal in a fireplace grate or even using a more efficient freestanding stove to capture more of the heat from combustion. Yet neither of those options were readily available in Boston or in other cities in the American North until well after the conclusion of the American Revolution. The lack of stoves is the most striking aspect of home heating during this period. Since Bostonians had struggled to secure firewood since the seventeenth century, the refusal to abandon the open fireplace and adopt a more efficient method of warming rooms seems very peculiar. No less a patriot than Benjamin Franklin had argued for using stoves or at least a modified version of one fitted into a fireplace, so why not adopt them?

Stoves did eventually make their way into urban hearths, but before this could occur, American society had to undergo an industrial revolution, perhaps as momentous as the political one served by the Liberty Tree. Although in theory the idea of replacing a fireplace with a stove was relatively simple—as Ben Franklin would readily argue—the actual way in which households began adopting stoves was the result of the interaction between much broader changes in the way goods were produced, distributed, and sold to customers. No single actor or institution coordinated these alterations, and yet the impact on home heating in American cities and beyond was undeniable. Stoves became the first important consumer durable good (one that is purchased rarely and does not wear out quickly) available in American marketplaces. Later in their history, Americans were quite comfortable with idea of purchasing large-ticket consumer durables like washing machines, automobiles, or computers. But in the early nineteenth century, the notion of a mass market for goods of this type was nonexistent. Developing an entirely new stove trade required iron makers to create these devices more efficiently and with an eye toward standardization, entrepreneurs to supply innovative designs that burned fuel more efficiently, and wholesale and retail agents to promote, display, and sell stoves with an eye toward cultivating a market for them. Cheap and effective stoves were the product of all these efforts, and they signaled the emergence of several interlaced industrial networks that promised a transformation of everyday life. Bostonians and their Liberty Tree are often credited with providing the spark that led to the American Revolution of the late eighteenth century; stoves played just as important a role in the industrial revolution that characterized the nineteenth century.[3]

The Fuel Crisis of the Early Republic

At the time of the Liberty Tree's demise, most American households burned firewood in fireplaces to stay warm. In the countryside, ready supplies of wood were hardly a problem; there the issue was finding the time and labor to convert timber stocks into usable fuel. Hardwood trees such as hickory, oak, and maple would have been the preferred choice in most fireplaces, as their density provided more efficient fuel and steady heat. Soft woods such as pine or spruce provided good kindling, but their high resin content and tendency to pop and crack made huddling up to the hearth less desirable. The standard measure of fuelwood is a cord, commonly accepted as a pile of stacked wood four feet wide, eight feet long, and four feet high. On farms, an able-bodied and skilled woodchopper could produce perhaps a cord per day, less competent farmers perhaps one-half to one-third of that amount. Since a cord provides 128 cubic feet of wood, moving and storing fuelwood was no small matter on American farms and rural villages. Houses needed space to maintain a woodlot for long-term fuel needs and a woodpile to store the cordwood and allow it to dry, and a significant part of daily labor was spent converting the cordwood into manageable sizes for the fireplace, usually twenty-four inches or less in length and sometimes with diameters of less than three inches. Estimates vary, but on average an American family in the northern climes required ten to fifteen cords of firewood annually in order to keep warm. Clearing forests for cordwood could also be profitable, with many farming families earning returns of up to three dollars per cord by selling surplus stocks of fuelwood in local markets.[4]

Urban residents faced a very different situation. Although large forests continued to dominate the interior of the nation, seaboard cities struggled from very early in their history to provide enough fuel to keep hearths warm. Boston, for example, encountered fuel shortages as early as 1637, and New York City drew upon the forests of New Jersey and Long Island for cordwood by the 1680s. Philadelphia fared somewhat better thanks to its nearby forests, but in 1706 the city hired an inspector to examine wood shipped to public wharfs, which suggests that Philadelphians needed influxes of fuel from the surrounding countryside. Urban Americans thus became reliant very early upon cordwood markets for their household fuel. Some of the participants in this spot market in firewood were farmers clearing the land and ridding themselves of excess cordwood. Others, however, sought to profit upon Boston's, New York's, and Philadelphia's lack of sufficient heating fuel locally. In 1680 New York's common council noted the "great Abuses and Injurys" that cordwood sellers inflicted on the city's residents "by the Sale of firewood by the Stick being of uncertaine and unequall Length, and Bigness." Complaints about unscrupulous dealers selling "short faggots" during the winter months continued to plague many seaboard cities for decades. Cordwood suffered from all the shortcomings of a spot market; gluts and shortages created volatile price swings, which often put buyers at a disadvantage.[5]

City authorities during the colonial era sought to regulate the cordwood trade by imposing strict regulations on the transportation and sale of firewood, especially as urban residents became more and more dependent upon fuelwood arriving by water. Prior to its occupation by British forces, New York City attempted to curb the frauds "daily Committed in the sale of firewood," as its Common Council reported in 1766. Three years later, New York set up a system of eight inspectors charged with ensuring the standardization of a cord and assigned them to particular slips along the waterfront. By the 1760s citizens in Philadelphia demanded the regulation of both boatmen and carters on the city's wharfs. Boston's regular shipments of firewood from Maine were essential to the city's survival; when the Port Act of 1774 relocated the flow of firewood to Marblehead, a concurrent spike in fuel prices added to the already high tensions in Boston. Colonial officials thus sought to ensure a steady supply of fuel to their cities, while also protecting consumers from rapacious dealers during the winter months of scarcity.[6]

The drive to regulate the firewood trade continued after the American Revolution, particularly as urban growth pushed the supply of wood fuel even farther from city centers. Contemporaries knew that future stocks of wood needed to be replenished, and one historian of American forests notes that by the end of the eighteenth century, the sale notices for farms highlighted the availability of woodlands for fencing and fuel for the first time. Replenishing the depleted stocks of wood in the Northeast became an issue of concern for both rural and urban residents. In 1791, for example, a Massachusetts farmer championed the idea of planting rows of trees as "screens or defences against the blasting winds" and to provide "nurseries for fuel." Philadelphia's *Weekly Magazine* tackled the issue with a series of articles in 1798 calling for public action. One writer suggested the formation of "a township forest, or free district of woodland" that would provide fuel resources for those in need, "particularly in regard to the poor, who even at this early day, procure it with difficulty, and at great expense." Another Philadelphian argued that the city should oversee the purchase of firewood upriver on the Delaware, ship chopped firewood down on rafts, and store thousands of cords in municipal wood yards for residents to draw upon when needed. Such planning would eliminate the "dissipation and thoughtlessness" that characterized the current trade.[7]

Rather than resort to publicly owned fuel markets, cities of the Early American Republic instead continued to regulate corders, carters, and other participants in the firewood trade, while leaving the supply of fuel to the whims of the free market. Mineral coal, although present in urban markets, was not yet a viable replacement for firewood, as the small and infrequent shipments of soft, or bituminous, coal from Great Britain and eastern Virginia could not satisfy the demand for domestic heating fuel. Instead, city officials tried to tighten up the existing fuel network. New York City retooled its firewood inspection system, creating a single inspector of firewood in 1802 and two years later dividing the city into fourteen inspection districts and requiring the inspector to report on the amount corded each month. New York's ubiquitous cartmen continued to be a vital link in the fuelwood economy. An 1814 law dictated exact rates for hauling firewood, increased the number of inspection districts to sixteen, and established a fine for anyone selling cordwood that did not meet official standards. Philadelphia's common council restricted the sale of cordwood to a small area in the city center to ensure uniform prices and allow for inspection. The city's sworn corders received an annual salary of $600, for which they enforced both size and length requirements for cords and ensured that no cordwood was resold within city limits between September and March. Although the reselling provision was lifted a few years later, municipal corders had significant control over the city's fuel supply; by 1810 they oversaw the unloading and sale of 68,691 cords at Philadelphia's public wharves. Local inspectors sometimes abused this privilege, as the New York carter John N. Johnson did when he was appointed firewood inspector for the West Side docks in 1816. For three years in this position, Johnson diverted the best New Jersey cordwood to his personal control, and by 1819 his fellow carters ran him out of office. Municipally controlled corders could ensure the uniform size of firewood and perhaps halt some of the most egregious gouging that went on during winter months, but they could do little to temper the wild swings inherent in firewood markets of the Early Republic.[8]

Any casual observer of America's urban landscape could see that fuel economy would be important to the nation's future, particularly as the dense spatial arrangement of cities and preferred construction style of dwellings made the conservation of heat difficult. As builders borrowed heavily from European styles developed in more temperate climates, American housing in cities like New York or Philadelphia remained notoriously inefficient at keeping out the elements. Although the Early Republic saw an increase in Georgian or Federal-style housing construction, in which high ceilings and white colonnades emulated British fashion, most city residences were squat, cottage-like wood dwellings that combined both work and living arrangements under one roof. As late as 1810, 65 percent of Philadelphia's housing stock was of wood frame construction. The city's distinctive "ground-rent" system encouraged builders to raise new structures quickly and without much thought to permanency, as they did not need to actually purchase the undeveloped land but instead assumed an annual rent payment that transferred to the purchaser once the house was sold. Structures both grand and humble employed fireplaces and chimneys for heat; wealthy residents were likely to have heating sources in nearly every room, while less affluent families usually made do with a single hearth. Some separation of domestic and working quarters occurred over the antebellum period, but the most likely arrangement of urban households would have seen a mix of family members, boarders, and tenants all working and living under a common roof and all tending to their respective fires individually. This urban landscape resisted any comprehensive or well-coordinated efforts to change the way households warmed themselves.[9]

Changes in fuel use needed to come from individual households, yet the most famous indigenous innovations available for American hearths—objects like Benjamin Franklin's Pennsylvanian stove—simply did not merit widespread adoption. Innovation in and of itself was not enough to bring about widespread change in home heating practices. Franklin himself admitted as much in 1786. While traveling back from Europe, Franklin revised the design of his apparatus after being inspired by a French philosopher's method of keeping smoke from billowing into the room—a common complaint surrounding the early versions of Franklin's stove. Franklin knew that his stove was "somewhat complex" and urged the beginner to "not be discouraged with the little accidents that may arise at first from his want of experience." Franklin asserted that "the studious man who is much in his chamber" would find his innovation most effective in heating a room. "To others who leave their fires to the care of ignorant servants," he continued, "I do not recommend it." The knowledge and skill necessary to warm a room with this improved design of the Pennsylvanian stove was more likely to fill the room with smoke than heat, so Franklin cautioned that "it is therefore by no means fit for common use in families." The Rittenhouse stove, developed by Pennsylvanian David Rittenhouse in 1784, attempted to improve upon the baffles of Franklin's 1744 stove with small iron plates that provided more radiant heat from a wood fire. Apparently this modification worked well enough in

Philadelphia, where a merchant house ordered 136 Rittenhouse stoves from a local ironworks. Elegant and expensive stoves following the Franklin or Rittenhouse design became more common in the parlors and drawing rooms of Philadelphia's wealthy households by the 1790s.[10]

These kinds of inventions, while appropriate for the drawing rooms of the elite, did little to alleviate the growing problem of fuel economy. In 1796, the American Philosophical Society offered a sixty-dollar prize for the best fireplace design for the "benefit of the poorer class of people, especially of such as live in towns, or other places where fuel is dear." The contest produced four entries that, emulating Count Rumford's theories, blamed large fireplaces and drafty chimneys as the main culprits in fuel wastage. "The same quantity of fire which warms the rooms of our richest people," one contestant noted, "leaves us to shiver on the hearths of our poorest." In 1799 the society awarded Charles Willson Peale and his son, Rafaelle Peale, the prize, even though their winning design essentially consisted of a Rumford fireplace with a sliding metal door installed at the front to regulate the draft of air to the fire and a damper in the rear to keep smoke from rushing back into the room. The Peales nonetheless patented their idea and announced that they would charge ten dollars for the installation of their fireplace. Both actions undermined the practical applicability of the Peale fireplace for less affluent households in Philadelphia and beyond. The elder Peale suggested as much in an 1803 article entitled "On Economy." Peale accused wealthy Philadelphians of being "afraid to make any changes, lest our servants not approve." Nonetheless, it was up to the city's betters to demonstrate fuel economy through example. "Since the imbecility of man renders example more of consequence than precept," he wrote, "let the wealthy begin to make the reform in the construction of their kitchens and other fire places." Although it generated little in the way of revolutionary innovations, the American Philosophical Society contest did signal recognition among its membership that improved fireplace designs should benefit the less affluent. The practical application of that principle, as Peale's attitude reveals, proved more difficult.[11]

If the complexity and expense of individual heating innovations ruled them out for the majority of households, what about building a massive hearth that could dispatch heat across a number of rooms? This idea has a long history. Roman engineers developed hypocausts, or heating systems in which central furnaces heated several chambers at once via an elaborate system of flues. Roman baths also heated water by sending it through thin brass pipes called "dracones" and then piping it directly to the pools for bathers. European systems of the eighteenth and nineteenth century mimicked the hypocausts without updating the design or the basic principle: burn large fires in a central furnace and force heat into various rooms via ducts or small pipes. Early heating engineers made a distinction between "direct" and "indirect" systems, with the former containing an actual heating element in each room and the latter providing heat by delivering heated air from a central location into the chamber. By the early nineteenth century, central heating systems using either hot air or steam were quite common in European public buildings, with steam generally considered the superior heating element, as it delivered more radiant heat across longer distances—that is, it more efficiently delivered heat to a number of rooms. Steam heat, though, required careful and constant management. One of the most commonly cited treatises on the subject in the nineteenth century, the Englishman Thomas Tredgold's *Principles of Warming and Ventilation*, lauded steam as the most efficient method of warming large buildings but stressed the need for experts to manage steam heating systems. Pipes and fittings leaked, and boilers exploded if not tended to correctly. "For, though in such hands it is perfectly safe and easily managed," Tredgold wrote of steam heat, "it is by far too complicated to be trusted in the hands of careless and ignorant people." Their complexity, coupled with the challenge of heating such large spaces with a single source of combustion, limited the use of these systems to public buildings during the early nineteenth century. Another English expert on the subject, Charles Hood, described the inadequacies of large-scale heating systems in his *Practical Treatise on Warming Buildings by Hot Water*. "The extremely unequal temperature of the flues causes an insurmountable objection to their general adoption," he maintained in 1844, "even if their great expense and difficulty of adaptation to dwelling-houses or public buildings did not operate against them."[12]

American designers like Benjamin Latrobe likewise experimented with hot air furnaces in the US House Chamber in Washington, DC, by 1808. New designs helped the almshouse and Pennsylvania Hospital warm the penniless and the infirm in Philadelphia. Daniel Pettibone, for example, designed a system that used a series of pipes designed to distribute air that was drawn from the outside, heated in a central furnace, and then distributed through charcoal-lined pipes that went from room to room. In 1810, the directors of the Pennsylvania Hospital claimed that Pettibone's system allowed them to use 75 pounds of firewood to heat an area that previously required 225 pounds. Pettibone also included testimony from Philadelphia's almshouse and the House of Employment in his promotional literature. Charles Willson Peale expanded upon his interest in home heating design by coming up with a heating system for large buildings with economy in mind. "However rich the man, and however plenty his fuel," Peale argued in 1798, "yet comfort and duty should induce him to embrace the means of economy here offered." Peale tested his design by building six improved fireplaces for New York City's almshouse. Even with these new central heating systems, though, individual rooms often employed fireplaces to supplement the furnace. Jacob Guild's Massachusetts Medical College, built in 1816, became the first building in the United States to be heated completely by a central furnace, which produced "a strong current of heated air . . . sufficient to warm the largest rooms in a very short time." Despite some exciting advances, then, innovations in central heating systems primarily benefited those on the extreme ends of the socioeconomic spectrum—large institutions serving the poor and wealthy homeowners who lived in large, multistory buildings.[13]

The Early Republic witnessed a burst of creativity in home heating methods from the nation's leading scientific minds. Although not all of them worked as well in practice as they did in theory, the proliferation of home heating refinements suggests a growing awareness of the problem. The limited diffusion of these improvements among the wider urban population, however, meant that the growing scarcity of firewood in those markets would continue unabated. As more and more land nearby was put into cultivation, moreover, stocks of available firewood dwindled. This had major implications for long-term prices of home heating fuel and made the likelihood of future energy crises even greater. Boston, for example, saw the price of hardwood more than double over the 1790s, and by the end of the decade some Bostonians paid as much as ten dollars a cord. One Philadelphia official estimated that even "genteel" families paid nearly $8 a cord, or $200 annually, to heat their homes. Volatile weather exacerbated the unpredictability of firewood markets, as uneven supplies in the best of times dwindled to practically nothing during periods of heavy snow and rain. During the winter of 1805 the price of oak wood in Philadelphia shot up to twelve dollars a cord—more than double its price earlier the same year. The *American Daily Advertiser* reported that one family, "having expended all their wood, was under the direful necessity, in order to keep themselves from perishing, to burn their table, washing-tub, and many other articles of household furniture." Many froze to death in their own homes. "Scenes like these proclaim the uncertainty of human possessions & enjoyments & the incapacity of man to protect himself against adversity," the Philadelphia merchant Thomas P. Cope wrote in his diary that winter. "It is during these seasons of suffering that the voice of omnipotence is heard & confessed." During the month of February 1809, Charles Peirce estimated the average temperature at twenty-six degrees and recorded three consecutive days in which the thermometer remained below zero. The poor suffered the brunt of Philadelphia's bitter season as firewood shortages caused the price of fuel to spike yet again. In this most temperate of the major northern cities of the Early Republic—where resident James Mease bragged, "Our winters are less uniformly cold, and more variable" and the Delaware River only froze solid for two weeks at a time by 1811—the pinch of cold weather coupled with fuel shortages to make urban life more and more difficult. Few residents were safe, as even the wealthy felt the bite of increasingly severe winters. New York City's well-heeled John Pintard wrote to his daughter in February 1817 that with the bitter cold he was "obliged to hold my pen to the fire to thaw the ink." "Indeed my ideas are almost congealed," he concluded.[14]

The Rise of the Stove

If more elaborate central heating systems and reconstructed fireplaces only appeared in the hearths of the affluent or institutionalized, a simpler or perhaps more realistic solution could be found in more simply constructed and modestly priced stoves. Of course, the Franklin and Rittenhouse designs became popularly known as "stoves," yet they both were really just improved fireplace designs and, as the inventors themselves noted, were hardly easy or inexpensive to install. A stand-alone stove—that is, a box made of iron or ceramic materials designed to lift the fire up from the fireplace and radiate heat more directly into the living space—had been long employed to heat rooms in central and eastern Europe. Stoves, both historical and modern, come in an endless variety of shapes and sizes, but the basic design helped differentiate a Franklin or Rittenhouse stove from its stand-alone relatives. Like the Franklin and Rittenhouse fireplace modifications, a simple five-plate or "jamb" stove—essentially an iron box stuck in a fireplace—offered radiant heat from its iron construction that was much more efficient than a roaring open fire. Eventually, though, "closed" stoves in which six cast iron plates constructed a box set on a frame with a pipe fitting to funnel smoke, became the more common design. With more iron plates, ten-plate stoves allowed cooking by creating a separate space within the simple six-sided iron box. Another simple cast-iron style was the "cannon stove," in which curved plates created a barrel-like chamber for a heating fire. These basic designs marked a simple yet profound break from the improved fireplace designs championed by inventors like Franklin, Rumford, Rittenhouse, and Peale. Many of them could be installed as stand-alone devices that only required carving a small hole in a wall or ceiling for the escape of smoke. Or stoves could be placed directly into large fireplaces and use the existing chimney for exhaust. However deployed, a box or cannon stove not only saved on fuel but could be easily disassembled, moved, and reassembled. These stoves thus provided portable heat, and their more compact iron firebox could help address the problems of both fuel scarcity and heat efficiency.[15]

Some early migrants to the American colonies brought their stoves and stove-making ways along with them. The immigrant communities of New Sweden and New Netherlands brought stoves from northern Europe. Germans also brought stoves with them when they sailed to Pennsylvania. Some German communities attempted to establish a stove industry in the American colonies. The most famous of these stove makers was the colorful "Baron" Henry William Stiegel, who immigrated to America from Cologne in 1750. Soon afterward he married Elizabeth Huber, whose father owned and operated one of the largest iron furnaces in Pennsylvania. By 1763, Stiegel and his business partners were running Elizabeth Furnace in Lancaster County and had acquired the Tulpehocken Hammer Forge in Berks County to manufacture cast-iron products. Stiegel renamed this complex Charming Forge and began making stove plates for six- and ten-plate stoves, sometimes stamped with the slogan "Baron Stiegel ist der Mann / Der die Oefen giesen Kann" (Baron Stiegel is the man / Who can make the stoves). Stiegel acquired a glassmaking business and continued to produce pig iron at Elizabeth Furnace, where he announced his arrivals and departures with a cannon shot from his custom-built seventy-five-foot tower. Stove makers rarely displayed such flair, yet Baron von Stiegel apparently made iron and glass better than he made loan payments; by 1775 Charming Forge was on the auction block, Stiegel was in debtors' prison, and his production of stoves had halted. Nevertheless, his brief success among Pennsylvania's large German population is an example of domestic stove manufacture in early America.[16]

Although stoves were being manufactured in the Mid-Atlantic region well before the Revolution, they were by no means common. One study of colonial home inventories found stoves present in only 3 percent of New York and New England households. In part, this is because of the location of production facilities. Eighteenth-century blast furnaces were placed in remote locations close to iron ore deposits and large stocks of wood for charcoaling. These "iron plantations" blended the iron ore, charcoal fuel, and a mediating agent, usually limestone, to make "pig iron," so named because when the furnace was tapped, the molten iron flowed into sand-lined depressions carved out on a casting-house floor that resembled (at least to the ironworkers) nursing piglets. This pig iron could be shaped into consumer products such as skillets, sash weights for windows, or stove plates either directly from the blast process or in a smaller furnace. The stoves plates and other products were made by directing molten iron into depressions made by intricate carved wood patterns—thus the name "cast iron." This is how the Baron's name appeared on his forge's stove plates; a more common pattern depicted a biblical scene such as the story of David and Goliath, the Annunciation, or the Last Supper. Since these iron castings were often made directly from the blast, they tended to be very thick and had many imperfections. In order to fabricate an entire jamb or closed stove from these plates, workers filed the castings to make a rough articulation that was snug but hardly airtight. These heavy, relatively expensive stoves made a striking but rare appearance in colonial homes.[17]

Although migrants from many northern European nations considered stoves essential to home heating, English colonists brought an entirely different sensibility to America. During the seventeenth and eighteenth centuries, an Anglo-American definition of comfort emerged that emphasized the need for circulated air and the ventilation of rooms. Conventional wisdom held that warm, airtight rooms would erode the inhabitants' constitutions and make them more susceptible to illness and disease. A roaring fireplace thus became both an aesthetic and a medical necessity. Strong drafts might chill to the bone as they delivered fresh air to a room; such was the price of good English health. "An Englishman may imitate the cautious habits of the people of colder climes," Thomas Tredgold's 1824 guide to home heating argued, "but he cannot change the variable nature of his own." Fireplaces provided notoriously uneven heat that could roast the occupants on one side of a room while leaving others freezing. Yet this tendency, too, was necessary for a good, strong English constitution. "It is certainly an overstrained idea of comfort to suppose an absolute equality of heat desirable," Tredgold argued. German families continued to find the stove an indispensable part of home life, even as they adopted Georgian exteriors to well-appointed houses during the Early Republic. Their stoves, however, remained tucked away in the interior of their homes and had little impact on the wider preferences for home heating. Despite North America's more severe swings in temperature, the cultural preference for open fireplaces and against closed stoves seemed to hold sway in most American households through the Revolutionary era and into the decades of the Early Republic.[18]

Preferences could change, however, as one correspondent to Philadelphia's *United States Gazette* argued in 1814. "Such is the force of habit, that in the north of Europe, where closed stoves are in universal use, the sight of a blazing fire is offensive to the inhabitants." "The predilection we feel in this country to sit around a sparkling fire during cold weather is," the author continued, "the result of habit. It is the province of rational beings, therefore, to extinguish it by degrees." But in order to become a more common fixture in American hearths, stoves required a more sophisticated network of manufacturers and marketers. Remote furnaces and forges could fabricate basic components, but they operated far from where stoves were needed most. Eighteenth-century iron plantations located in the mountains of Pennsylvania, Maryland, and Virginia required easily accessible and abundant supplies of ore and wood for charcoaling. Facilities known as forges or foundries that made cast-iron goods from pig iron were less limited by location. Urban forges and foundries did become important producers of cast-iron consumer products in the years after the War of 1812, but they still needed to secure pig iron from furnaces located in remote areas, and it was cheaper to make stove plates directly from the blast. The average antebellum furnace used the equivalent of 150 to 500 acres of wood a year, depending upon efficiency. A small furnace might therefore require about three thousand acres of reserve forest, assuming a twenty-year regrowth of fuel stock. Large operations like Pennsylvania's Hopewell Furnace needed five or six thousand cords of wood each year to create about one thousand tons of pig iron. Their voracious appetite for charcoal and ore meant that blast furnaces needed to be close to those resources, usually far from cleared and populated areas. The best way to ship heavy goods like stove plates was via waterways, but the distant location of many iron-making facilities ruled out that mode of transportation.

Most stoves traveled to local markets by some combination of horse-drawn wagon and water transport. Martha Furnace, Jesse Evans's operation in the Pine Barrens of southern New Jersey, manufactured pig iron, hollowware, and stove plate for customers in the immediate area, including Philadelphia. Stoves were a regular part of Martha Furnace's business, showing up in terse entries in the clerk's diary such as "Mr. Fugery brot his oval pattern for 9 plate stoves," "Luker & Cox carted stoves to the Landing," and "Mick making feet for stoves" in the summer of 1810. Martha Furnace demonstrated the typical characteristics of the early stove trade: it serviced and relied upon local markets for the most part, offered few changes in design from year to year, and did not attempt to specialize in stove manufacture.[19]

Broader changes in the American economy after the War of 1812, most notably the improvement of turnpikes and canals and the advent of railroads, transformed the nation's iron trade. The drive to cut costs and increase the speed of transportation through the construction of land and water routes between cities involved massive amounts of capital and construction on an unprecedented scale. The most famous of these projects was the Erie Canal, an ambitious system that used dams, locks, aqueducts, and other cutting-edge transportation technology to create a 350-mile water link between Albany and Buffalo. The Erie Canal came with an astonishing $7 million price tag—a figure that in today's dollars would be measured in billions—and was financed, constructed, and operated by the State of New York. After its completion in 1825, the cost of shipping goods from Buffalo to New York City fell by roughly 90 percent. New York's success story inspired other states to build canals; Pennsylvania, Ohio, Indiana, Illinois, and Michigan all devised expensive and publically funded canal systems during the "canal boom" of the 1820s and 1830s. Many of these projects failed to earn a profit, and by the 1840s and 1850s the heyday of state-funded internal improvements had passed. The subsequent development of the nation's rail network came mostly through private investment, but these overland routes blended with the waterways constructed with public funds to create a dense network of new transportation options for iron manufacturers. When the cost of transport dropped, so did prices. By 1860, the cost of land transport of heavy items like iron had been reduced by 95 percent from that of the 1810s. Although much of this came as the result of private initiative, it is important to remember the role that state authorities played in kick-starting this revolution in transportation. As public and private projects raced to connect markets, the cumulative impact on American manufacturing was profound. For the iron trade, it meant that bulky commodities like stove plates could be shipped across greater distances and at less cost than ever before.[20]

Some small furnaces across the North continued to manufacture stove plates for local markets, but centralization of the iron industry accelerated in response to this improved transportation system. The appearance of urban forges and foundries, many of them drawing their regular supply of pig iron from newly constructed canals, helped move the production process closer to the retail customer. The proprietors of blast furnaces usually acted as wholesalers in the trade, selling stove plates directly to retail outfitters, who assembled and sold them. As more Americans became acclimated to the use of stoves and their improved fuel economy, specialized "jobbers" facilitated the flow of iron from rural furnaces to large markets like New York, Philadelphia, and Baltimore. Dedicated stove manufactures, particularly those operating smaller foundries, also began to serve the growing demand. By the 1820s, it was common to see stove plates cast in both rural blast furnaces and urban foundries in retail markets, with the actual assembly, smoothing, and finishing of stoves often taking place in close proximity to urban consumers. Philadelphia, for example, emerged as a vibrant center for the financing and selling of stoves, even though the city did not serve as the sole location for the manufacture of all stove components. Major furnace operators kept offices in Philadelphia, with more and more stoves assembled in the city. One of these furnace owners, Samuel Gardiner Wright, invested in the Delaware and Dover furnaces in southern New Jersey, even as he lived and worked in Philadelphia. From there, Wright developed relationships with retailers as far away as Massachusetts and Maryland and helped build the urban market for stoves. One Salem, Massachusetts, merchant asked to see Wright's patterns "of such kinds as you think would answer in this new market." A decade later, retailers were coming to Wright in Philadelphia. In 1831 John Bradbury & Company of Newburyport, Massachusetts, announced to Wright that they would be ordering ten to twenty tons of stove plate and that they intended "to visit Philadelphia early in the season for the selection of patterns." The city's location straddling the Schuylkill and Delaware Rivers, allowing for water transport to both the state's interior and the entire Eastern Seaboard, helped establish Philadelphia as an important center of the stove trade. Carting stove plates over rough-hewn roads in the countryside bit deeply into the producer's cost; shipping them via waterway made much more sense.[21]

As stove manufacturing began to flourish with distinct nodes of production and consumption during the 1820s and 1830s, the trade also saw a burst of inventive activity and a rush to secure patents on new designs for stoves. The US Patent Office issued 53 patents between 1790 and 1815. Over the next fifteen years, government officials issued 329 patents in this area, and by 1835 1 in 10 patents issued dealt with stoves. This rush for patents represented a dramatic departure, as early stove and fireplace designs provoked little competition. Rittenhouse's slight improvement upon the Franklin stove or the various American adaptations of a Rumford fireplace were considered by contemporaries to be a public service. Franklin never sought to patent his stove design and quite literally gave the idea away to a friend. Iron molders guarded early stove patterns, but they seemed so rudimentary that claiming a unique provenance over their intellectual design would appear ridiculous, like owning the blueprint of a frying pan or a sash weight. As the consumer demand for stoves increased during the 1820s and 1830s, though, improvements to the basic six-, eight-, or ten-plate stove became quite common. From the simple box design of the Revolutionary era, when heating rooms was the device's only function, nineteenth-century stoves blossomed to serve cooking, baking, and decorative functions. Newfangled grates improved the intensity of the flame, innovative plate designs radiated heat from the firebox, and decorative columns drew smoke away from living and working spaces. Consumers could purchase stoves with three or four legs, busts of famous Americans like George Washington, and varying amounts of decorative trim. As the variety of stoves in the marketplace increased, so did the incentive to protect innovations. The huge size of urban markets in the United States and a political climate that encouraged inventiveness frustrated attempts at controlling the pace of innovations or restraining individual furnaces, forges, or foundries from manufacturing virtually any kind of stove they found profitable. The American system of granting patents, moreover, demanded only a relatively small filing fee of thirty dollars and offered a simple administrative process open to inventors at all levels. This effectively democratized innovation by making patent protection available to a wide range of manufacturers, both small and large. Design innovation became the province of stove manufacturers, not professional inventors or scientists, who were likely to seek only one or two patents over the course of their lifetime. These stove makers often noted which patent they used, along with lengthy descriptions of their devices, directly in their advertisements and trade catalogs (fig. 1.1).[22]

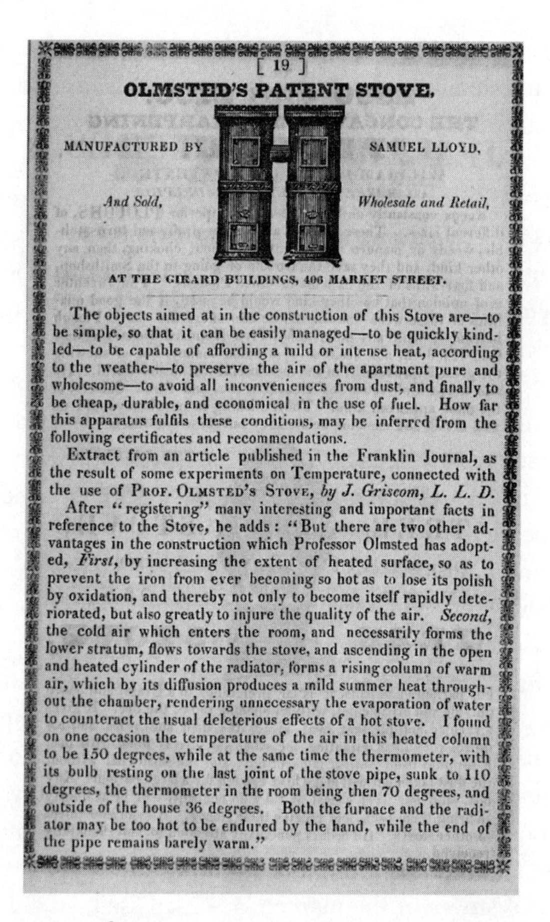
[19]

OLMSTED'S PATENT STOVE.

MANUFACTURED BY SAMUEL LLOYD,

And Sold, Wholesale and Retail,

AT THE GIRARD BUILDINGS, 406 MARKET STREET.

The objects aimed at in the construction of this Stove are—to be simple, so that it can be easily managed—to be quickly kindled—to be capable of affording a mild or intense heat, according to the weather—to preserve the air of the apartment pure and wholesome—to avoid all inconveniences from dust, and finally to be cheap, durable, and economical in the use of fuel. How far this apparatus fulfils these conditions, may be inferred from the following certificates and recommendations.

Extract from an article published in the Franklin Journal, as the result of some experiments on Temperature, connected with the use of PROF. OLMSTED'S STOVE, *by J. Griscom, L. L. D.*

After "registering" many interesting and important facts in reference to the Stove, he adds: "But there are two other advantages in the construction which Professor Olmsted has adopted, *First,* by increasing the extent of heated surface, so as to prevent the iron from ever becoming so hot as to lose its polish by oxidation, and thereby not only to become itself rapidly deteriorated, but also greatly to injure the quality of the air. *Second,* the cold air which enters the room, and necessarily forms the lower stratum, flows towards the stove, and ascending in the open and heated cylinder of the radiator, forms a rising column of warm air, which by its diffusion produces a mild summer heat throughout the chamber, rendering unnecessary the evaporation of water to counteract the usual deleterious effects of a hot stove. I found on one occasion the temperature of the air in this heated column to be 150 degrees, while at the same time the thermometer, with its bulb resting on the last joint of the stove pipe, sunk to 110 degrees, the thermometer in the room being then 70 degrees, and outside of the house 36 degrees. Both the furnace and the radiator may be too hot to be endured by the hand, while the end of the pipe remains barely warm."

Figure 1.1. Advertisement for Olmsted's Patent Stove, 1840. Early advertisements for stoves highlighted their novelty and design, often with mention of a patent. Patents were difficult to enforce in reality but offered the promise of something new and innovative to the prospective buyer. Stove makers also included testimonials from satisfied customers and scientific authorities to demonstrate the cutting-edge nature of their technology. *A. M'Elroy's Philadelphia Directory for 1840* (Philadelphia: Isaac Ashmead, 1840), 19

•

The story of how three individual stove makers helped create urban markets for stoves demonstrates the relationships between innovation, production, and marketing. First, James Wilson of New York was a stove manufacturer who engaged in invention throughout his career. Wilson sold thousands of stoves over three decades, but his experience in the business demonstrated the pitfalls of trying to retain a proprietary stake in stove design. In 1816,

Wilson obtained an affidavit in Poughkeepsie, New York, claiming that he alone was the "true and original inventor or discoverer of the invention or improvement of the Franklin Stove" for "the purposes of heating Air and cooking." He forwarded the affidavit along with a petition to President James Madison asking for exclusive rights to use the patent. Wilson found that asserting, much less protecting, this claim was difficult, but he continued to sell stoves for the next two decades. Both the open system of granting patents as well as the difficulty of enforcing them meant that innovations in stoves circulated without much oversight. Enforcing a patent was up to the holder, who could rack up huge legal bills without much success. By the 1830s, Wilson had moved to New York City, where he still made stoves, still took out affidavits insisting that he was the true inventor of innovations such as the "pyramid hot air double grate anthracite coal stove and feeder for heating rooms," and complained when he saw others stealing what he claimed were his inventions. When he spotted the delivery of a familiar-looking stove on the street one day in 1839, he rattled off a letter to its sellers, Morse & Son. "I have now been 26 years in the Stove Business and [am] the only one now in the Same Business that Commenced when I did," Wilson wrote. "I have expended thousands of dollars in making Improvements in all kinds of Stoves," he continued. Wilson had secured patents on stove designs, but that hardly thwarted rival manufacturers and dealers. By 1842, even Wilson's attorney gave up on his client's claims of provenance, informing the veteran stove manufacturer that his designs were too basic to secure a patent, much less prevent any rival dealers from selling similar-looking stoves.[23]

Wilson ran afoul of another influential stove inventor from Schenectady, New York, Eliphalet Nott. Whereas Wilson represented the practical side of stove innovation, Nott took the same intellectual approach to heating as a Benjamin Franklin or Count Rumford. Nott served as the long-time president of Union College, where he first began experimenting with stoves in 1815. Although his academic predecessors like Franklin believed that their improvements in home heating should be disseminated widely, Nott sought to protect his intellectual property through the US Patent Office. He secured a patent for his "Fire-Place and Chimney" in 1819 and received three more in 1826. Nott was a bit more successful at defending his inventions; when Wilson complained in 1833 that he had stolen one of his designs, Nott actually won the lawsuit and secured damages from Wilson. Nott turned to large-scale manufacturing when he opened the Union Furnace up the Hudson

River in Albany. Run by his son, Howard Nott, the Union Furnace was a foundry capable of turning out a thousand tons of cast iron annually. Although it was not exclusively dedicated to stoves, this facility served as one of the area's leading stove manufacturers until it was wiped out in the Panic of 1837. Before it left the stage, however, Eliphalet Nott's concern demonstrated a more winning combination in the stove making trade: innovation *and* production.[24]

Jordan Mott of New York City invented stoves as well, but like Professor Nott's, his pioneering work on the production process and his plan for reaching urban consumers were more significant to the overall development of the trade. If Wilson's case highlights the role of innovation and Nott's the significance of production, then Jordan Mott enhanced the marketing techniques of the stove trade by interacting with customers more regularly and tweaking the designs and functions of his stoves to meet those needs. Mott's new approach had personal origins, as he began his career not as an ironworker or furnace manager but as a grocer. Drawing upon this everyday experience with consumers, Mott became a leading retailer in stoves at time when the business was becoming much more specialized. When Mott began selling stoves, most hardware dealers tried to hedge their investment in stoves by selling them alongside other iron cast ware. For example, only two of Providence, Rhode Island's, six hardware stores sold stoves in 1824. Eight years later, Providence boasted a full-time stove dealer and manufacturer as well as twelve hardware stores selling stoves. In 1815, a New York City directory listed only two stove manufacturers and retailers, Broadway's Birdsall & Heafield and Water Street's Gilbert Brown. Fifteen years later, fourteen stove businesses were listed, with three other stove manufacturers joining Brown & Company's expanded works on Water Street. Mott honed his expertise in the stove trade from his own shop in the Water Street stove district, where he and other retailers would get the stove plates shipped from blast furnaces in New Jersey or Pennsylvania, assemble them in their small workshops, and then sell them to the public. Mott was good at design innovation; his 1833 patent for a self-feeding hopper solved the problem of using smaller amounts of fuel in stoves, and he continued to secure more patents and win medals for his inventions.[25]

Mott's innovations were popular with his customers; he bridged the gap between invention and marketing masterfully. For example, Eliphalet Nott found a way to use Pennsylvania anthracite coal in his stoves. This was a significant scientific breakthrough, as anthracite had been dismissed by some critics as a "rock, not coal" because of its difficult ignition—more on that in chapter 2. Mott tinkered with the idea of putting mineral fuel into his self-feeding hopper, drawing cool air into a wide firebox base, and then circulating hot air out into the room. The "base-burning" stove was a major success, as the rush by other stove makers to copy Mott's innovation proved. In fact, Mott's creativity helped sell stoves out of his shop, but as fellow New Yorker James Wilson demonstrated, patents were notoriously difficult to enforce, particularly in a fiercely competitive market. By 1845, there were sixty-six stove manufacturers and dealers in New York City, with twenty on Water Street alone. The area became a center of stove design and assembly (fig. 1.2). Mott became a leader of this trade through a simple innovation in the production process that helped move stove making away from remote blast furnaces and, in turn, allowed the industry to concentrate production in the Empire State. More specifically, it was Mott's utilization of the cupola furnace that changed the pattern of stove production. Cupolas are smaller, cylindrical furnaces that use lower temperatures to melt scrap and pig iron into casted forms. Because they do not require the intense heat of the smelting process, cupolas can operate with much less fuel and labor than blast furnaces. Prefabricated cupolas can also be shipped anywhere and require relatively modest amounts of iron and fuel for their operation; this made them ideal for urban manufacturers like Mott, who could use the cupolas to control costs and thus survive in the increasingly competitive consumer markets emerging in American cities.[26]

The entrepreneurial efforts of Wilson, Nott, and Mott helped make stoves efficient and cheap; the growth of American cities and the nation's internal improvements network accelerated this trend. The next generation of stove makers seized upon the use of cupola furnaces to make Albany and Troy into new centers of a dynamic stove trade. Although Philadelphia emerged as an early node as the industry began to move away from its decentralized roots in the rural iron plantations, upstate New York became the center of America's stove industry largely as a result of foundries' using cupola furnaces. Following the opening of the Erie Canal in 1825, Albany and Troy straddled the crossroads between the well-established urban markets of the Eastern Seaboard and emerging ones in the western United States. This region had a thriving iron-casting industry before the advent of stove foundries, but the arrival of the stove industry in the late 1820s complemented Albany's existing

Figure 1.2. An "Air Tight" stove design from a Water Street stove maker, 1840s. Stoves sold in New York City came in all shapes and sizes. The presence or absence of decorative trim often reveals whether stoves were meant for wealthy or less affluent customers. This "air tight" model came with the kind of elaborate ironwork that suggests it was designed to fit into a well-appointed home. *Documents of the Assembly of the State of New York, Seventy-Second Session*, vol. 7 (Albany: Weed, Parsons, 1849), 122

industrial sector perfectly and took full advantage of its transportation facilities. Joel Rathbone's company, for example, maintained sales offices in Albany as well as a large foundry along the banks of the Hudson. Rathbone's company became the largest iron-making firm dedicated to stove production by 1844, largely by streamlining the stove-making process. Once pig iron arrived via the Hudson River, Rathbone's molders used a cupola furnace to melt it down and poured the molten iron into a two-sided flask made from wooden frames and filled with molding sand. Flask casting could be done at blast furnaces, but it required a higher degree of skill and talent than the preferred method of sand casting. Once the iron castings cooled in the flask, they were removed and went to the next stage of production, finishing, where rough spots in the plate were filed down to ensure a good fit; then the parts were passed on to the stove mounters, who assembled the final product. Firms like Ransom and Rathbone dealt in high-quality stoves, which demanded both tight fits and good-looking exteriors. By the advent of the Civil War, Rathbone's foundry used three thousand flasks to produce up to thirty-five thousand stoves annually. Their stoves could be sent cheaply to markets throughout the United States and beyond via the Erie Canal and Hudson River shipping networks. "A modern traveler," J. Leander Bishop wrote in 1864, "asserts that he saw the Rathbone stoves in Constantinople, and on boats far up the Nile."[27]

By midcentury, Albany and Troy each boasted seven foundries employing fifteen hundred workers making seventy-five thousand stoves annually that were worth roughly $2 million. The concentration of Albany's foundries mirrors the development of the trade in general, as stove making had become a thoroughly urbanized industry by 1850. Albany and Troy accounted for a large share of the $6 million worth of stoves manufactured in 1850, but stove foundries popped up in Providence, Buffalo, Pittsburgh, Cincinnati, Louisville, and Saint Louis in the antebellum period. New York City and Philadelphia remained important in stove manufacturing; by the 1850s Philadelphia had five large manufactories dedicated to stove production, as well as a host of smaller foundries that made plates and other stove accessories. Innovations like the use of cupola furnaces shifted the manufacture of stoves from remote blast furnaces to urban centers, thus making stoves nearer both to the primary consumer market in cities and to the network of roads and waterways that could cheaply transport them to smaller markets. American cities became critical centers for both the production and consumption of stoves. Urban dealers placed their wares on display, and stoves became a common sight on the streets of nearly every American city. Figure 1.3 depicts "Foering & Thudiums Cheap Stove Ware-House," one of Philadelphia's most prominent stove manufacturers and retailers, as its storefront on North Second Street appeared in 1846.[28]

Amid all this entrepreneurial innovation and change in home heating markets, what happened to the old cultural bias against stoves? Once the economic side of the stove network took shape, Americans' reservations about these heating implements became less formidable. Customers could

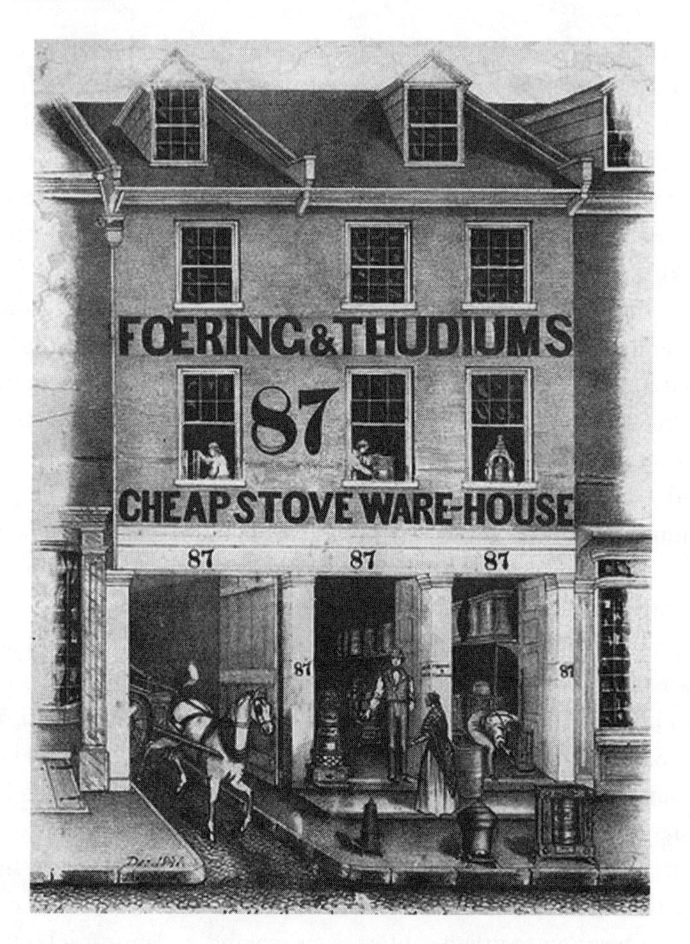

Figure 1.3. A Philadelphia stove dealer, 1846. Stove makers displayed a wide range of models for their customers, as demonstrated by the variety of stoves visible in this advertisement. Note the proud declaration that this shop sold "cheap" stoves. By the 1830s and 1840s, the stove dealer was a common fixture in the commercial districts of American cities. Library Company of Philadelphia

now choose from a wide range of stoves at various prices, which undoubtedly helped dispel their bad reputation. In fact, stoves seemed thoroughly modern to them. Among the last holdouts of this cultural bias were New England's venerable churches. Tales of an epic battle between stove and antistove factions in the churches of New England gained popularity over time. By the time many of these stories made their way into print, the antistove faction stood as an easy metaphor for tough old New England customs and their resistance to change. The apocryphal tale, for example, of antistove worshippers being driven out into the fresh air by the fumes emanating from the new stove only to find that there was no fire present pops up in several reminiscences of the old days in New England. Several versions of this story were told over time. Even P. T. Barnum—no stranger to hokum—related the tale in his 1855 autobiography. Ever the showman, Barnum doubled the number of ladies overcome by imaginary stove fumes and made them "venerable maidens" suffering from the "dry atmosphere and sickly sensation." Yet, even as the particulars of the stove/antistove fights might be more legend than fact, the stove's delayed appearance in New England's churches marks the stove as a new technology that some Americans regarded with some degree of uneasiness. Barnum, for example, describes the prostove faction as "quite ahead of the age" while ridiculing the old deacon who complained that the new stove drove the cold air to the rear pews, "making them three times as cold as they were before."[29]

Stoves made it possible for more and more Americans to live, work, and even worship in relative comfort. As the industry consolidated in American cities and producers like Jordan Mott competed in an increasingly competitive environment, stove prices generally came down. Of course, this market was segmented, and an affluent consumer could always purchase a high-quality, top-end stove that combined efficiency and style. More important, though, stoves became affordable for middling to poor consumers over the 1830s and 1840s. The incredible variety of stoves makes it difficult to estimate exactly how much the "average" stove dropped in price over this period. The average cost of a ten-plate stove, for example, dropped by one-half in real prices, from forty dollars in 1810 to twenty dollars in 1852. Trade catalogs and newspaper advertisements list prices as high as fifty dollars in the 1810s and 1820s for a high-end stove. By the 1830s and 1840s many of the same models had dropped to thirty dollars. For the average American family, this was still a fairly high price, which is why stoves are considered the first real consumer durable. Creative stove retailers also pioneered ways to rent and sell secondhand stoves during this period, which made them widely available to consumers with no credit and little cash. The increasing affordability of stoves widened the ability to heat urban households in the decades following the War of 1812.[30]

Stoves played an important role in expanding available methods of home heating and also offered one potential way to alleviate the intermittent supplies of cordwood that plagued the growth of cities of the Early American Republic. As stove manufacturers and retailers expanded their base of operations, their product became a more familiar part of the urban household. Though the roaring open fireplace was an important cultural element of Anglo-American home heating at one time, by the 1820s and 1830s firewood shortages made this method of warming buildings prohibitively expensive for most urban households. But even if the purchase, lease, or inheritance of a stove helped warm a home, what could a family do amid the chronic fuel shortages that seemed to appear with more and more frequency each winter? The arrival of stoves solved one part of the home heating equation by supplanting a very familiar technological system with a new industrial one, but it did not completely dispel the problems of fuel supply that plagued American cities.

By the 1840s and 1850s, many stoves burned anthracite or bituminous coal instead of firewood. Yet the transition from wood to mineral fuel was by no means an easy or obvious choice for Americans. The rise of coal as a significant fuel in home heating required the collective efforts of individual entrepreneurs and consumers but also drew upon unlikely agents such as philanthropic organizations and canal companies to become a real alternative to cordwood. The campaign to replace wood with coal was by no means coherent or centrally planned, yet the successful adaptation of coal to the nineteenth-century hearth was a significant change in the history of home heating and, perhaps more important, resulted in the creation of a massive system of production, distribution, and consumption that connected urban hearths directly to the emerging industrial economy.

2 How Mineral Heat Came to American Cities

PHILADELPHIA'S ELIZA LESLIE knew the value of a good and economical fire. As the author of several books on cooking and domestic economy in a literary career that stretched from the 1820s through the 1850s, Leslie helped shape the character of the middle-class American home in the antebellum decades. In her 1840 volume entitled *The House Book; or, A Manual of Domestic Economy*, Leslie devoted an entire section to "Fuel, Fires, &c.," in which she outlined the methods of using stoves, the character of various types of wood, and how to save money on fuel. In the latter category, anthracite coal played a prominent role in Leslie's advice. "In buying anthracite coal, (as in most other things,) that of the best quality is eventually the cheapest," she wrote. "It goes further, lasts longer, gives out more heat, with less waste from slate-stones and ashes, and leaves better cinders when it is extinguished; and good cinders may always be turned to account by burning them over again." But anthracite took some skill to burn effectively. Known colloquially as "hard coal," anthracite was notoriously difficult to ignite and could go out unexpectedly. "A skillful firemaker will fit in the large and small pieces, so as to consume both to advantage," Leslie asserted. "We have known this done by servants who took great pride in the excellence of their parlour fires; for instance, a coloured man, who always assorted his coal, and brought it up separately in two scuttles, reserving his finest pieces for the front of the fire, and calling them his *facers*." Once expertise with "hard coal" had been achieved, fuel savings inevitably followed, as did a steadier and more accommodating flame. "Anthracite fires, if managed *exactly* according to the preceding directions," she concluded, "will be found more comfortable, more economical, handsomer in appearance, and in every respect more satisfactory than if conducted in any other manner. This we know by experience."[1]

Eliza Leslie summarized a fairly complex process in only a few short pages. That her "hard coal" provided such a steady and efficient flame was the result of ancient photosynthesis—or the conversion of solar into chemical energy by plants—being trapped by layers of sediment and rock over millions of years. That burning coal releases the stored energy of ancient forests and earns it the name "fossil fuel"—this mineral actually came from organic origins. In the nineteenth century, Americans knew that their country held enormous reserves of coal, which broke down into two main categories. Bituminous, or "soft" coal, replaced wood very easily, as it burned—and smoked—in almost the same way as wood. Throw a piece of soft coal on a fireplace, millions of Britons could attest, and it will burn readily. This is because bituminous coal has a lower percentage of carbon, usually ranging from 50 to 85 percent. The lower carbon content makes bituminous coal easier to ignite, but its high level of impurities creates a dense smoke and soot. Another thing that Britons could attest to was the unpleasant byproduct of bituminous coal fires; the choking "fogs" in large cities like London were a daily reminder of it. A denser, but rarer, mineral with higher proportions of carbon—usually around 90 percent—than the more common bituminous stocks, anthracite fires could go out quickly if not tended properly. A specific skill set, then, was necessary in order to use anthracite for domestic or manufacturing purposes. In the early years of its use, many found anthracite impossible to burn, hence its somewhat derisive moniker "stone coal." And yet Pennsylvanians boasted of the vast quantities of anthracite tucked into three small but rich coalfields in the eastern mountains of their state. If only their anthracite could come into common use, the benefits of this mineral fuel had the potential to transform everyday life; certainly burning coal in lieu of firewood could help alleviate the devastating fuel shortages that seemed to plague cities of the Early American Republic with alarming frequency.[2]

But this substitution was not easy. As the history of the adoption of stoves proves, a well-established everyday practice is difficult to replace. Generations upon generations of humans had burned organic fuel in the form of firewood. Demonstrations of how coal burned with more intensity and economy were nice for scientific circles and salons, but a sustained transition from organic to mineral fuel required a sustained transformation of everyday household practices on par with the most radical changes that the Industrial Revolution brought to the workplace. The collective impact of thousands of households making the decision to adopt coal in lieu of wood eventually led to the widespread use of mineral fuel. In fact, the seeds of fossil fuel dependence that plagues twenty-first-century Americans can be found in the hearths of the early nineteenth century.[3]

This chapter focuses on the transition from organic to mineral fuel in two major American cities in the decades following the War of 1812. As a large city that suffered from endemic fuel shortages, Philadelphia seemed most ripe for some alternative form of heating fuel. The close proximity of the city to northeastern Pennsylvania's abundant anthracite reserves made it the most likely candidate to undergo a rapid transition from wood to mineral fuel. As it turns out, though, Philadelphians were introduced to coal use in a number of different ways, depending upon their economic status. Wealthy and poor Philadelphians found distinct ways of negotiating changes in heating technology, which delayed the wholesale adoption of coal for two decades. By the 1830s, when anthracite made the leap from its most proximate city to the nation's largest market, New York City, the pattern of selling coal to a wide range of consumers had been well established. Changes in home heating thus continued to blaze new paths for the American economy; if stoves became the nation's first consumer durables and were indicative of the nation's industrial prowess, then anthracite served as the first major American innovation in the consumption of mineral energy. It was no mistake, then, that Eliza Leslie declared that "we know by experience" about anthracite. Just how this transition occurred, and perhaps more important, *who* accelerated the use of coal are interesting questions. Wealthy consumers may have understood that anthracite saved money, but who convinced them of that? How did they, or their servants, learn how to use anthracite in stoves or fireplaces? Did this change in fuel use spread from wealthy households to less affluent ones, as did the use of stoves? How exactly did the transition from organic to mineral fuels occur in everyday life?

Early Failures in the Coal Trade

A quick look at a map of coal reserves in the United States shows that the nation contains an incredible amount of coal; in fact it contains roughly one-quarter of the world's supply. Much of this coal lay within striking distance of the Eastern Seaboard (fig. 2.1). Late twentieth-century pundits, amid a menacing oil crisis, dubbed America the "Saudi Arabia of coal." But it is important to remember that abundance does not necessarily translate into widespread use, as many nations blessed with copious reserves of mineral wealth struggle to employ it to their benefit. The nation's coal trade in the late eighteenth and early nineteenth centuries did not look all that promising. Two early coal regions, one in the South and one in New England, failed to cultivate fuel markets. But even as they floundered, these two regions helped create the foundations of a national market for coal by providing three basic elements that would define the trade for the nineteenth century. Mining interests in those regions first agitated for a federal protective tariff that would shelter the nascent coal trade. Early coal miners also highlighted the need for cheap and efficient transportation in their trade; their failures signaled the critical need to get coal from the mine to the hearth as quickly and as cheaply as possible. Finally, and most important to the story of home heating, early boosters of the coal trade knew that domestic markets—the adoption of mineral fuel in urban households across the nation—were of vital importance to the industry's future. Each of these factors relied upon a series of political and economic networks that blossomed in the decades after the War of 1812. Like the increased use of stoves, the adoption of mineral fuel depended on the development of early industrial networks; without them, American coal remained virtually useless as a source of heating fuel.

The shortcomings of the early American coal trade offer an excellent example of how entrepreneurial talent and initiative alone could not kick-start the transition from organic to mineral fuel. The first major coal-mining region in the United States was in eastern Virginia in a small bituminous coalfield outside of the capital city of Richmond. The proprietors of coal mines there—called "colliers" after the English fashion—first began digging for coal in large open pits before the American Revolution; by the outbreak of the War of 1812 Virginia colliers exhausted the easily extracted reserves and resorted to British-style shaft mines that plunged hundreds of feet below the surface. Mine owners adopted the "breast and pillar" method of mining,

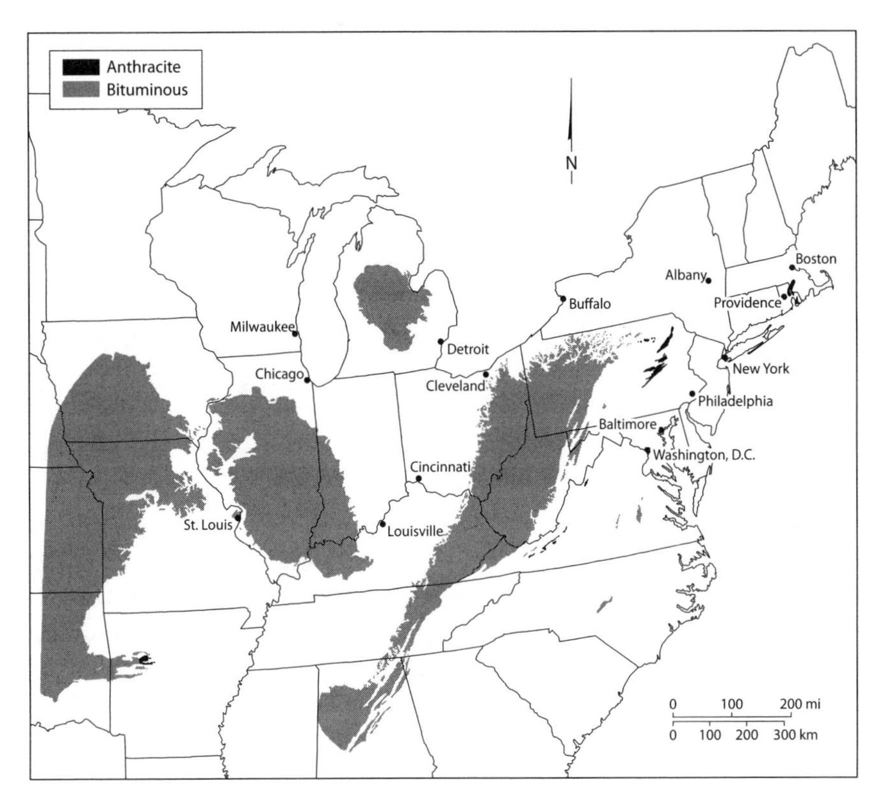

Figure 2.1. Coal fields in the northeastern United States. Bill Nelson

•

in which they excavated large chambers of coal from the seam—called the "breast" in miner's parlance—and left support pillars of coal to keep the ceiling in place. Using explosive charges of black powder to loosen the coal, miners shoveled it into carts that mules pulled up to the surface. It was a rather unsophisticated system but the most effective one for the time. Virginia's mixed workforce of free white and enslaved and free black miners raised about fifty thousand tons of coal annually by the War of 1812, and Richmond-area colliers hoped that as soon as the British blockade was lifted by either victory or treaty, they could ship their coal to the fuel-hungry cities of the Northeast. Ambitious colliers like Chesterfield County's Harry Heth sought to adopt cutting-edge technology in their mines, cultivated wholesale and retail contacts in northern cities, and invested thousands of dollars in coal mining. Heth, for example, estimated that he had sunk $30,000 into his Black

Heath mines by 1814 and predicted that by the end of the war they would "be noted of inestimable value." The Richmond Basin colliers did not lack for optimism, entrepreneurial talent, and a sharp understanding for what they needed in order to grow their trade.[4]

Early in the region's history, for example, Virginia colliers complained about British imports of coal cutting into their business and sought political solutions to the problem. Vessels carrying coal, usually as ballast, from English ports had been selling mineral fuel in American markets for decades. Local merchants referred to it as "sea coal," or "Liverpool coal," not because it was mined in those locations but after its place of origin. In 1798 twenty-one Virginia colliers petitioned Congress to raise duties because they could not "support a competition in the market with foreign coal" coming from the Atlantic. Although a sustained flow of coal from Europe to the United States was not likely, the threat of foreign coal was enough to flood the nascent market for mineral fuel at a particularly vulnerable time in the American coal trade's development. "That the quantity of coal, daily raised, greatly exceeds the demand," the Virginia colliers argued, "and although one hundred and fifty coasting vessels are now occasionally employed in the coal trade, a very great surplus of coal remains undisposed of—circumstances which evince the reasonableness of an expectation that this trade may be made productive of national benefit." Congress did not respond to this initial demand, arguing that high duties "tend exorbitantly to increase the price of this necessary article" and that a protective tariff might "deprive some portions of the Union of their necessary supply."[5]

Fortunately for the Virginia colliers, federal policy makers eventually agreed that mineral fuel was a commodity deserving of a protective tariff. The original 1789 tariff on coal imports stood at two cents per bushel, a measure of volume that, in the case of coal, usually weighed about sixty to eighty pounds. Duties increased gradually over the years until in 1812 it reached ten cents a bushel, or about 15 percent of the wholesale price of British coal. The tariff on coal dropped to five cents a bushel in 1816, which ranged between 10 and 25 percent of the price of coal in New York, and remained at about that percentage until 1842. British imports bounced back after 1815, but they never again exceeded more than 10 percent of American production. Of course, American colliers constantly complained about competition from abroad, but domestic production dominated fuel markets, and the United States became a net exporter of coal in the 1870s. Political actions at the federal level helped shield American colliers from significant international competition throughout the antebellum period, but protective tariffs were not enough of a boost to establish Virginia as an important center of the nation's coal trade.[6]

Colliers in eastern Virginia also realized the importance of transportation to their business, but local geography worked against them even as they enjoyed national demand. Their coalfield straddled the James River about ten miles upstream from Richmond's tidewater port, but the relatively short journey from the mines to coastal vessels proved notoriously difficult for colliers like Harry Heth. Turnpike companies were loath to allow coal-carrying wagons on their right of way, as the heavily laden carts created deep ruts in the road. The major water thoroughfare for the area, the James River Canal, lay on the opposite side of the river from Heth's mines, and so he was forced to load and unload his coal via wheelbarrow a number of times in order to reach a canal boat. Each time workers transferred the cargo they added to its cost, and their shovels broke the soft bituminous coal into small bits. Once the coal arrived at Rockets, Richmond's deepwater port, laborers transferred it to a coastal vessel that could take it to Heth's many retail contacts in Philadelphia, New York, or Boston. The lack of an efficient transport system reduced the quality of Richmond Basin coal and raised its price; soon it developed a bad reputation among northern merchants. As a result, the expected postwar boom never materialized. In 1817, for example, Harry Heth expected to sell a large amount of coal in the growing fuel markets of the Northeast. "The case however is not so—I have at the time upwards of 200,000 bushels on hand & not a solitary order from any of the great towns of the North & East as usual at this season of the year," he wrote to a business associate. "On the contrary, all my letters state, that they are supplied with foreign coal on much better terms than I can possibly afford it." Virginia's agriculturally minded politicians showed little interest in subsidizing their state's coal trade, and so the much-needed water or rail systems that could efficiently transport coal from the mines of the Richmond Basin to the city's tidewater port stalled in the legislature. Without those important transportation networks in place, Virginia's coal could not take advantage of the protection offered them by the federal government.[7]

What about colliers working in close proximity to urban markets? Could small distances allow mineral fuel to thrive there? The case of Rhode Island anthracite offers an opportunity to explore that question. "The inhabitants of the eastern States, and of Boston in particular," one 1808 pamphlet proclaimed, "are much indebted to the genius and perseverance of one of their own countrymen, for the discovery of a Coal Mine at Rhode Island, so eligibly situated in every respect, that a constant and regular supply can be obtained at all times, without difficulty." Rhode Island's anthracite enjoyed several geographical advantages over Virginia's bituminous coal in serving the fuel needs of New England; coal in that region lay practically on the shoreline, where it was a short journey for a vessel to Boston or any other towns that needed mineral fuel. Because New Englanders were "so long accustomed to the use of wood for fuel," the 1808 commentator conceded, "it is not surprising that much ignorance should prevail on the subject." But once the "the introduction of a more oeconomical, and, in many respects, a more convenient substitute" took hold, Rhode Island anthracite appeared to have a very bright future. By 1814, the Rhode Island Coal Company (RICC) sought to seize this initiative. After receiving a small infusion of capital via a state-sponsored lottery, the directors of the RICC went about the hard work of convincing New Englanders to burn their coal. "Long use must of course beget an attachment to the usual materials of fuel," they admitted, "and we look up on wood as a sort of household god." But, they continued, "we may look forward to no very distant day, when wood, as an article of fuel, will become a luxury." Their mine was "near at hand," their supply "inexhaustible" and much cheaper than Virginia or Liverpool coal. The RICC knew that anthracite posed a challenge to the uninitiated user. Its prospectus, not surprisingly, then, attached a guide for ignition along with the standard testimonials touting Rhode Island anthracite's economy and utility. Since anthracite was notoriously difficult to light, very explicit directions concerning the laying of wood kindling, the separation of the burning coal from its ashes, and the need for a steady supply of fresh air were necessary to help potential customers learn to use "stone coal."[8]

But even close proximity to Boston, a city that had struggled to secure firewood for centuries, could not overcome problems of production and the public's lack of familiarity in the Rhode Island coal trade. Although mining engineer and coal trade historian Howard Eavenson has estimated that the Rhode Island coalfields were capable of providing between five and ten thousand tons of anthracite annually during the antebellum period, the RICC struggled to ensure a steady flow of coal out of its mines. Correspondence from the RICC's agent, Samuel Waldron of Portsmouth, suggests that the company had a large amount of coal ready to ship. But here, as in Virginia, local practices worked against the ability of New England colliers to exploit opportunities created by protective tariffs. Rhode Island colliers measured their product in "chaldrons," which was an English standard of volume, not weight. Like the bushel standard used by Virginia colliers, a chaldron could vary widely depending upon the size, density, and chemical character of the coal. Bushels ranged anywhere from sixty to eighty pounds on average; chaldrons were between two and three thousand pounds. Shipping by volume, not weight, made for uneven quantities, which merchants did not appreciate. (The 2,000 pound "short" or 2,240 pound "long" ton later became the American coal industry standard.) Waldron, moreover, had to contract with captains in the open market to secure shipping. In 1823, a manager at the RICC's mines, Thomas Cary, wrote that he "was somewhat sorrowful about the coal," as he could not secure a vessel to transport it up to Boston. The RICC struggled with rates and reluctant captains throughout the 1820s and 1830s, as well as confusion over exactly what amounts they were shipping. When H. D. Sedgwick tried to secure a customer for RICC coal, he wrote to Waldron about the confusion in "considering the difference between the ton by which Lehigh Coal is computed & the chaldron." Without reliable supplies and transport, then, bushels of Virginia bituminous coal and chaldrons of Rhode Island anthracite became increasingly marginalized in early American fuel markets.[9]

The RICC admitted in its 1814 prospectus that there were problems. "With regard for this coal, it is true, that it cannot be burned in a common fireplace, instead of wood,—the only test that many have ever made of it; nor can it be easily burned in a grate of common construction," they conceded, "nor can it, without considerable labour, be ignited in the same manner, as the more bituminous coal; and therefore, as some conclude, it won't burn at all." By the 1820s, this rather modest self-criticism had become an accurate prophecy, as Bostonians preferred coal from regions other than Rhode Island. A booster of the RICC chided New England consumers for spurning the local trade: "You are calculated upon as consumers of the Lackawaxan coal, after having been borne for hundreds of miles on the waters of Pennsylvania and New-York, and paid tribute to your canals. . . . If you will consent to this," the diatribe continued, "it is to be presumed that you will have no objection to send to Pittsburgh for your iron castings." A few years later H. D. Sedgwick rejoiced at the "general introduction of stoves for the purpose of burning

Anthracite" in Boston but bemoaned the poor place of Rhode Island coal in that city's fuel markets. "I am aware that Rhode Island Coal has been a proverb, a by-word, and a reproach," he wrote. "It is has exercised the wit of some good jokers, and a great many poor ones." Sedgwick hoped that "with moderate means" and "competent skill" the RICC could be revived to serve as "a source of private emolument and public advantage," but the battle was already lost. Even the stalwart Samuel Waldron, who devoted over two decades of his life to the cause of Rhode Island coal, saw its ultimate dismissal by the 1830s. In the summer of 1835 his business partner in New York City, John Rynex, described the coal market there, now dominated by the superior "Lehigh and Lackawanna" anthracite from Pennsylvania and concluded that "at Rhode Island mines there are large quantities [of coal] that lie useless—it is so."[10]

So why are the short-term failures of the early Virginia and Rhode Island coal trade important to the history of the transition from organic to mineral energy? Their problems suggest that there were some key elements to success in delivering mineral fuel to urban hearths: protective tariffs, cheap transportation, consumers' familiarity with the product, and a steady supply for receptive consumer markets. The presence of coal itself was irrelevant without a full complement of the components that make up a mineral fuel network. Both the Virginia and Rhode Island coalfields enjoyed one or two of these advantages early in their history but could not establish all of the elements that could make their mineral fuel cheap and easy for urban consumers to use. These early miners did not have networks in place that could maintain enough consumption of their product to create more demand. They were early movers in the industry, to be sure, but since Virginia and Rhode Island colliers tried to tap into urban home heating markets before established networks for carrying and marketing mineral fuel had taken hold of American cities, they remained on the margins of the nation's coal trade.

The Rise of Anthracite Coal

Instead of Virginia or Rhode Island, Pennsylvania became the center of the mineral fuel revolution in home heating. Pennsylvania's "stone coal" was abundant, although the three major anthracite regions were located in a remote and sparsely populated sector of the state. The shipment of coal overland from the anthracite regions to eastern cities was simply not feasible, and consumers' unfamiliarity with "stone coal" presented another problem.

The story of Pennsylvania anthracite might have gone the way of Virginia or Rhode Island coal, but colliers in the Keystone State worked in a very different context than their competitors. The Schuylkill Navigation Company (SNC) and the Lehigh Coal and Navigation Company (LCNC) both used a combination of improved waterways and canals to provide direct links between Pennsylvania's anthracite coalfields and urban centers. The LCNC completed enough of its line in 1820 to send 325 tons to market; the SNC shipped 5,306 tons in its inaugural year of 1825. These two companies constantly promoted the quality of "Schuylkill" and "Lehigh" anthracite (the generic names of coal shipped down each line) and competed for market share in Philadelphia and points beyond. Although the LCNC's charter allowed the company to own and operate coal mines in addition to maintaining its canal and the legislature restricted the SNC to carrying only, both were mainly "navigation" companies, and thus operated in an incentive structure that privileged toll revenues over price manipulation. In other words, two competing firms that could both mine and transport coal might collude in order to artificially raise the price of coal prices at market by withholding shipment; the uneven distribution of privileges in the early anthracite trade made that coordination impossible, and once the leaders of the SNC decided to encourage as much traffic on their route as possible, the LCNC could only respond in kind. Withholding coal from urban markets in order to increase prices simply wouldn't work for the LCNC, as the SNC would be happy to carry the lion's share of coal to market. As a result, the strategies of both the SNC and the LCNC ensured that the volume of "stone coal" arriving in Philadelphia increased dramatically throughout the boom years of the 1820s and 1830s as prices generally dropped and anthracite carriers took a sharp interest in promoting its use.[11]

Anthracite boosters in Pennsylvania, like the colliers in Virginia and Rhode Island, knew that they needed to cultivate customers in order to ensure their own future. In this case, though, the promoters of coal were not individual proprietors but well-funded and socially connected corporate interests. Both the SNC and the LCNC were authorized by the legislature of Pennsylvania to improve water navigation on their respective rivers; the responsibility of constructing those links and, perhaps more significantly, ensuring steady traffic along those completed lines, lay with the companies themselves. Therefore, the cause of anthracite became one and the same with the future prospects of the SNC and the LCNC. The president of the

SNC, Cadwalader Evans, articulated the vision of coal's potential impact on large urban markets in an 1817 public address to Philadelphians. Evans sold firewood before he became involved in the affairs of the SNC, so he was well aware of the problem facing urban fuel markets. As president of the SNC he also had an obligation to increase the potential market for anthracite and used his position as an influential and well-connected member of Philadelphia's business community to do so. Evans stated that his firm could ship coal down the Schuylkill River to the city at a cost of thirty cents a bushel—by his estimation this was the equivalent of reducing the cost of oak to three dollars a cord. He immediately brought poor families into the equation by exclaiming: "What a relief this would be to the indigent!" "Remember, fellow citizens, how much you subscribed and paid last winter to supply the poor with wood," Evans exhorted. "All the country too near the Schuylkill which is bare of timber, and all the towns on that river, and on all the tide waters of the Delaware will reap the benefit of this reduction in the price of fuel." Evans and other anthracite boosters pushed their product on all ranks of Philadelphians, of course, but at a very early point they also suggested that coal would serve as the "workingman's fuel" because of its efficiency and abundance.[12]

Coal made its way into upper- and middle-class hearths in a roundabout fashion, as the conversion from organic to mineral fuel did not occur overnight. Anecdotal evidence suggests that elite households led the way in providing testimonials that appeared in both newspapers and promotional literature. For example, in an 1825 letter appended to a LCNC promotional pamphlet, Matthew Carey, a prominent publisher and public figure in Philadelphia, noted that he had "for three years, used Lehigh coal in my study, and found so many advantages in it, that I would not exchange it for hickory wood, even if I could procure the latter gratis." He claimed that there were three reasons for his conversion. First, he cited "an entire security from the destructive wants of fire, unless in cases where there is the most wanton carelessness." Second, the heat seemed more uniform. "I used to be occasionally very uncomfortable, even with considerable fire burning," Carey wrote, "and my room, as you know, is by no means large. My feet used to be cold almost always at night, in winter. Since I have used this coal those grievances are removed entirely." The final reason is that when he burned coal, he found that the fire burned longer and without constant repair—he found his room "comfortable on rising," with the thermometer at 60 to 62. This last endorsement brought economy into the mix, as burning coal was cheaper than wood. Carey noted that his son-in-law's kitchen wood fire cost thirty-seven and a half cents per day, but two coal fires, one of which was kept burning constantly, cost only twenty-five cents a day.[13]

Even though he cited the economic value of coal, Carey's main emphasis was on the *comfort* provided by anthracite heat. The whole idea of comfort was a relatively new concept in the early nineteenth century, and in elite circles its rise merged with notions of upper-class respectability. The idea of creating a comfortable home with a coal fire, then, became a way for Philadelphians to either reinforce or announce their higher standing in society. Eyewitness accounts from elites testifying to this new home heating fuel's quality became a common theme in the promotional literature for anthracite. Because skepticism abounded as to anthracite's combustibility and the pejorative "stone coal" label still clung to it, public exhibitions aimed at affluent audiences were meant to reinforce written testimonials from social peers. Jesse Fell of Wilkes-Barre, one of the first promoters of anthracite, recognized the importance of firsthand accounts of anthracite's flame. "On its being put into operation," he remembered in 1826, "my neighbours flocked to see the novelty; but many would not believe the fact until convinced by occular demonstration." Coal companies sponsored these demonstrations in well-heeled hotels, often aimed as much at soliciting stock subscriptions as gaining actual customers. Personal testimonials from prominent individuals also aimed to convince consumers that anthracite was worth the effort of conversion. Promoters hoped word of mouth would continue to spread the word about anthracite. In 1859 Sidney George Fisher, for example, recalled how oral testimony convinced him that anthracite was the best means of building a comfortable fire. Not everyone in the older generation of Americans was persuaded, but coal seemed to be the fuel of the future:

> I remember when coal was first used. Long after, hot-air furnaces were invented, which are now universal. I recollect, many years ago, 20 I suppose, I one day advised old Genl. Forman to get a grate for his parlor, telling him of the trouble saved & that by burning coal the room could be made comfortable in any part of it. His niece, Miss Augusta Forman from one of the lower counties, was present, and supported my opinion, saying that one of her neighbors had got one and that actually the water *never froze* on the sideboard the whole winter! The old General, however, preferred the wood fires, which are indeed much the pleasanter, if you can warm a house with

> them. The true way in the country is to have a furnace for the hall & wood fires in the rooms.

The diffusion of information about anthracite is difficult to track in quantitative terms, but it is easy to find evidence that among the upper echelons of Philadelphia society, anthracite became a stylish home heating fuel.[14]

Coal became an important part of the urban landscape in only a few decades. In 1831, Thomas Porter updated the guidebook *Picture of Philadelphia* to reflect the many changes in the city since James Mease published the initial version two decades earlier. Porter noted that "coal was but partially in use" in 1811, but it was in common use by the time of his writing. "The advantages that we reap from those concerned in the coal trade," he elaborated, "is of considerable moment," and low fuel prices had "become almost a drug." Porter noted the eighty-one thousand tons of coal that came down the Schuylkill in 1831 and estimated that the mining, shipping, and transporting businesses were worth about $500,000. Anthracite sold for between five and seven dollars a ton at that time, but there was every reason to expect the price to fall to four dollars per ton in the near future. Porter's description might warm the hearts of Philadelphia's coal dealers, investors, and customers, but the city's residents still burned a considerable amount of firewood. According to estimates from *Niles' Register*, firewood still accounted for 66.5 percent of Philadelphians' fuel expenditures in 1830, as opposed to 33.5 percent for coal. Three years later, at the height of the anthracite boom, coal's share had increased to only 35.3 percent of the more than $1.14 million spent on warming Philadelphia. How could the rest of the city's hearths be converted from wood to mineral fuel?[15]

What the early coal trade needed was a wide base of consumers—drawn from both the wealthy and the poor—to purchase mineral fuel and sustain growth. The process of "consumption," or the purchase of goods in a marketplace with a range of real choices, was an essential component of American capitalism in its formative years and beyond. Consumers had the power to change long-standing assumptions about the structure of American society as early as the American Revolution, when colonists united to enforce "nonimportation" as a means of political resistance against British policies. As the United States struggled to make its own way after independence, consumption shifted to serve as an important benchmark of economic prosperity and self-reliance. For individuals, participation in a growing marketplace of goods could provide markers of status and affluence without signaling an abandonment of the republican principles required by the new political order. A warm room or house was one of the many markers of a comfortable life, and as the industrial economy of the United States grew in scope and in size, the marketplace would grow apace to satisfy the consumer's expression of individualism.[16]

A choice of consumer goods, however, does not necessarily mean an equality of access to them. Scholars who look at the history of consumption describe how it tends to first mark and then transcend social hierarchies as the consumer habits of the wealthy are eventually taken up by lower-income groups. Items that are at first limited to the wealthy eventually work their way into broader-based markets. Merchants aggressively supported these changes in the economic landscape, as it widened their customer base and increased profits. Sugar and confectionary goods, for example, shed their distinction as luxuries as they became more affordable to a broader consumer base. Consumer markets eroded definitions of elitism based on birth or marriage as they offered middling and even poor folks the ability to emulate elite lifestyles. This "consumer revolution" thus had the power to shake out long-standing social structures and reshape them with new markers of status.[17]

Even if coal did serve as a better substitute for firewood, its potential consumers required a new skill set in order to burn it for heat. Specific equipment and knowledge of anthracite's properties were required to convert existing fireplaces, and even among elite buyers, a cultural preference for open fireplaces slowed the adoption of coal stoves and total replacement of firewood. The purchase of coal meant that mineral fuel technology needed to be adopted in each individual hearth, and the technical barriers to burning hard coal manifested themselves not in the marketplace but in the home. Historians of technology often differentiate between the *consumers* of a technology and its *users*. These individuals might be one and the same, but sometimes those who use a particular technological system did not originally purchase it. In many nineteenth-century households adhering to proscriptive roles for men and women, the difference between consumers and users broke down along gendered lines. The expectation was that men would be the consumers but women the users of mineral fuel. This particular division of labor in urban households kept many women from formally participating in the marketplace, yet their value in adopting new technologies like coal-burning stoves or fireplace grates was essential to the rise of mineral fuel specifically and to industrial development generally. Women in most households were not paid to maintain the hearth, nor would they necessarily have seen the public demonstrations of anthracite's effectiveness. They often found themselves users of a technology whether or not they were the ones who purchased it; they had to learn how to use coal in order to connect their households to the coal trade. In 1849, for example, the Boston merchant Supply Clap Thwing supplied a family friend, Susan Ela of Concord, New Hampshire, with anthracite. Along with an explanation that his price of $6.50 a ton was quite reasonable and likely less than a local dealer would charge, Thwing gave explicit instructions on how to manage an anthracite fire. "The grate should be cleaned out every morning with the fire built up fresh," he wrote. "In letting down the ashes there will be a good deal of unburned coal. This should be carefully lifted out either daily or weekly & put on the fire to avoid waste." Thwing was teaching Susan to be both a user and consumer of coal. The trick of extending this process to broad swaths of American households was a major challenge.[18]

The lack of technical knowledge required to use coal was telling among both wealthy and poor urban residents. Problems with ignition persisted even among those who were in domestic service to elite families that made the conversion from firewood to coal. "Very few servants at first understand the method of kindling and continuing a fire of Lehigh coal," one house servant wrote in an 1827 domestic manual. "Many will never learn, and many more from erroneous instructions, whilst they think they understand it, make but a bungling piece of work of it." Rather than constant manipulation, burning anthracite required patience and a steady supply of fuel. One expert complained, "Servants never learn this mystery, they always fly to the poker in every case of distress, and by their stupid use of it, double their own labor and vex the mistress of the house." Eliza Leslie warned homemakers to instruct their servants in the art of making a coal fire and warned against "injudicious poking and stirring" that would put out the fire instead of improving it. Denison Olmsted argued in 1836 that disseminating the skill of anthracite combustion would be a "top-down" process. He wanted to "diffuse among the more intelligent portions of the community, knowledge of the *principles* on which the most successful management of coal fires depends. . . . They must condescend, next, to superintend personally, the construction and regulation of their fires," he continued, "until their domestics are furnished with the necessary practical skill, which they will acquire much sooner by example, than by verbal or written instructions."[19]

The struggles of users to burn coal in the American household is illustrated vividly in David Claypoole Johnston's 1832 cartoon entitled "Anti-Phlogistic" (fig. 2.2). As one of his generation's great satirists, Johnston took on a myriad of causes in his illustrations. "Anti-Phlogistic" depicts the an upper-class gentleman's attempts to have his servants build a fire with anthracite (referred to as Rhode Island coal after that state's small anthracite field) in his fireplace. Although one of the well-heeled observers proclaims that "you'll find it a very economical kind of fuel" and the other announces that it is "likewise a very safe kind of coal," the coal stays unlit despite the best efforts of the two servants. The angry experimenter proclaims anthracite to be a very "innocent, inoffensive, kind of coal" and that, in "a general conflagration of things," Rhode Island would be the last state to burn. It is difficult to discern what the particular problem is in "Anti-Phlogistic," and the title's invocation of the mysterious substance of "phlogiston," considered by premodern chemists to be a material substance that provided heat, suggests that anthracite ignition remained inscrutable. The presence of bellows suggests that the servants were making the common mistake of blowing too hard on the fire. Blacksmiths familiar with "stone coal" knew that one needed to pile on wood kindling, leave the fire be, and then be certain that the fire remained separated from the ashes. Blowing or poking the fire could actually put it out. As one guide noted, "Failure most frequently proceeds from stinginess in the material of kindling," and "you will never have a good anthracite fire, till you have broken your husband, a brother, or wife, of the mischievous habit of poking." The same pamphlet waxed eloquently about anthracite's "steadiness, constancy, and trust-worthiness. Like a man of integrity and consistency, you always know where to find it. It plays you no tricks, but maintains the same sober, equal demeanor." Soon the method of igniting anthracite seemed like common knowledge, even when this technical knowledge was expressed in paternalistic or even condescending tones. "Well—step by step, improvement was made on improvement and this coal, that ignited with so much difficulty," the editors of Baltimore's *Niles' Register* argued in 1834, "became a familiar or household fuel—the most stupid easily kindling a fire with it, assisted by small portions of wood."[20]

Figure 2.2. David Claypoole Johnston, "Anti Phlogistic," in David Claypoole Johnston, *Scraps No. 3 for 1832* (Boston, 1832). This satirical print highlights the difficulties of igniting anthracite coal and the struggles of both consumers and users to burn "stone coal." Phlogiston was a mysterious substance thought by eighteenth-century scientists to cause combustion, so labeling the cartoon "Anti Phlogistic" parodied the problems that many early adopters found with anthracite coal. An explanation of the dialogue at the top of the print is found in the text on p. 55. Library Company of Philadelphia

A Benevolent Heat?

By the 1830s the spread of effective practices for burning anthracite to poorer households came from a perhaps unexpected source. Philanthropic organizations, which combined financial and charitable impulses to push anthracite on their clients, connected the drive for wider coal consumption with a humanitarian desire to keep the poor warm. Philadelphia's Fuel Savings Society appointed a committee to explore the distribution of coal along with firewood in the summer of 1831 at the urging of its secretary, Lindzey Nicholson. Although wood had the advantage of familiarity and immediate value, they found the adoption of coal would be of great benefit "to the *laboring classes* of our citizens, *in particular*." Indeed, they estimated coal would save families approximately six dollars per year on fuel. Nicholson not only wanted to reduce fuel costs for the poor, as a member of the board of managers of the Schuylkill Navigation Company, he also had a personal stake in anthracite's widespread use.[21] The Union Benevolent Association (UBA), another Philadelphia philanthropic organization, founded in 1831, hoped to "elevate and better the condition of the poor by inculcating the principles of an efficient morality, and calling forth, or cherishing in their minds, a spirit of independence and self-estimation which will produce habits of thoughtfulness and reliance on their own resources." More specifically, W. H. Keating, a member of the UBA executive board, proposed that the visitors they sent to poor families should suggest that they purchase a "proper stove or grate for anthracite." Keating's endorsement of anthracite came as no surprise, as he had delivered a landmark address touting the future of the American coal trade to the American Philosophical Society a decade earlier. Later in life, Keating would become one of the founders of a major anthracite carrier, the Philadelphia and Reading Railroad. In addition to endorsing Keating's plan, the UBA also tinkered with the idea of converting coal dust—a waste product in increasing abundance during the 1830s—into a cheap fuel for distribution to the poor. These programs attempted to kill two birds with one stone by increasing consumers for anthracite while stemming the immediate suffering caused by endemic fuel shortages.[22]

Convincing the poor to switch to anthracite required more than financial incentives. A consumer's conversion to the new fuel took periodic investments in brand new technology. Wealthy philanthropists probably had personal familiarity with this adaptation but from a vastly different perspective than their clients. How did wealthy consumers convert to anthracite? Philadelphia's city treasurer, Cornelius Stevenson, carefully recorded all of his fuel expenditures from 1831 to 1842 in his receipt book. This decade-long run of purchases reveals a few trends among affluent consumers of fuel. First, there is a consistent investment in new technology. In 1832, for example, Stevenson paid $57.86 to install a coal grate in his kitchen and the following year had a coal cellar installed beneath his house. In 1839, he paid nearly $200 for a brand new furnace. Over the course of the decade, Stevenson spent $375.86 on various improvements to his home, most of them related to coal-burning technology. Second, Stevenson also never stopped purchasing wood, which suggests that his household never made a complete transition from one fuel to the other. Wealthy households enjoyed the luxury of using different fuels, perhaps installing a fireplace grate in one room, buying a coal stove for another, and retaining a traditional fireplace in yet another. Third, the timing of the coal purchases reveals an important pattern of consumption. Once Stevenson had a coal cellar installed, he was able to purchase large amounts of coal, around ten to fifteen tons at a time, to secure fuel throughout the winter season and sometimes into the next year. Among many wealthy customers, a single annual purchase of fuel was common. A long-range strategy of consumption, not an option in the hand-to-mouth reality of many poor families, meant that coal purchased cheaply in the summer could be stored and might last through two winter seasons. The ability of wealthy consumers to lay in a supply of coal thus guarded them against rising winter prices and potential disruptions in supply—a strategy that the managers of fuel philanthropies would have known well and most likely practiced themselves by the 1830s.[23]

Constraints on space and budget made it difficult for working families to emulate middle- and upper-class fuel use on their own, so philanthropic societies attempted to bridge the gap for coal consumption among their clients. In Philadelphia they underwrote the cost of purchasing stoves in order to subsidize the up-front investment required to burn anthracite and enlisted canal corporations in the campaign. One visitor to a poor household in January 1831 noted the absence of a fireplace or stove and reported that the family warmed itself by burning charcoal in a floor "furnace" and letting the smoke go out the open window. Several coal-related remedies for the poor emerged that winter. The Lehigh Coal and Navigation Company advertised a $1.50 anthracite cooking stove as providing "economy and solid comfort for the poor." That same year, the Fuel Savings Society resolved to purchase one hundred stoves from Steinhaur & Kisterbock and sell them to depositors at $5.50 each. The Union Benevolent Association went even further, distributing between 350 and 400 stoves for cooking, heating, and baking to poor families. The stoves were stamped "UBA" on the side, lest anyone try to resell them. This might seem a modest program, given that Priscilla Clement estimates that 11,060 individual Philadelphians (over one-tenth of the entire population) received some form of charity in 1830. But as a parallel to the public trials in affluent hotels, the UBA loan offered one way to promote anthracite consumption among Philadelphia's working poor and reach an often overlooked class of customer. Overall, these programs developed into a substantial subsidy for new fuel technology, as cheap coal stoves sold for about fifteen to twenty dollars during the 1830s, with most models averaging about thirty dollars. Although no documentation survives, one assumes that UBA agents also instructed their clients on the vagaries of anthracite ignition. In this case, then, the aims of philanthropy and the market merged to encourage technological change in Philadelphia's poorer households.[24]

Despite the efforts of corporate and philanthropic actors, the conversion to anthracite among poor consumers took some time. In November 1831, the Fuel Savings Society purchased ninety tons of coal but could not find enough poor families who could burn it. The following winter, the Union Benevolent Association reported that 4,562 persons—over 40 percent of all welfare recipients in Philadelphia—asked for firewood, not coal. Five years later, the directors wrote, "Wood is an article of great and indispensable necessity, and a large part of our means is expended in its purchase," at the same time that they were distributing the UBA coal stoves to encourage the use of mineral fuel. Evidence suggests that the balance tipped in the favor of coal sometime in the 1840s. In 1849 the UBA sold 960 tons of coal and 138 cords of wood at half cost. Most likely the limitations of space and income restricted options for poor consumers. The conversion to coal use was, moreover, an all-or-nothing decision in a household with a single stove or fireplace. In this light, the preference for firewood among the poor reported by the UBA or Fuel Savings Society—often in patronizing terms that assumed ignorance or closed-mindedness on the part of their clients—might have revealed the desire for less affluent consumers to keep their options open or to have someone else pay the cost of making the transition to coal.[25]

The introduction of anthracite to America's largest consumer market, New York City, developed in a similar fashion. Canal companies led the way in initial promotion, and a broad campaign to promote "stone coal" hit consumers at all levels. The saga that unfolded when the Delaware and Hudson Canal Company (D&H) sought to introduce anthracite in that city saw another deliberate blurring of philanthropic and profit-oriented agendas in which corporate interests took the lead. The D&H was the brainchild of William and Maurice Wurts, two brothers from Philadelphia who had made a small fortune in the dry goods trade and eventually accumulated coal lands in Carbondale, Pennsylvania, as a settlement of debts with a local landowner in the Lackawanna mining region. The Wurts brothers recognized the potential of anthracite but found breaking into the Philadelphia market difficult. Instead they turned their focus to New York City. Their attempt to capture heating fuel markets there hinged upon the futures of two firms: the D&H, which would form an all-water route from Honesdale, Pennsylvania, to the Hudson River, and a client coal-mining firm called the Lackawaxen Coal Mine and Navigation Company. By 1824 the Wurts brothers were ready to open stock subscriptions to both firms and made appeals to investors based upon generous estimates of profits. They proclaimed, for example, that they could deliver anthracite to New York City at a cost of $2.64½ cents per ton and sell it at $6.00 a ton—a price that would undercut both British coal and other suppliers of Pennsylvania anthracite. In turn, with their estimates of an annual demand of 160,000 tons in the New York area, the Wurts brother promised profits of up to 35 percent on their initial capitalization of $1.5 million. They announced that stock subscriptions would open both in Manhattan and at locations along the canal route in January 1825.[26]

The D&H's Lackawanna coal, as it turned out, burned brightly in the eyes of investors and residents along its line but initially less so in the eyes of New York City's consumers. New York's *Commercial Advertiser* noted that although anthracite could be used for some purposes, for domestic heating British or Virginia bituminous coal made a "more lively and cheerful fire." After the D&H began shipping anthracite to New York in earnest in the season of 1828–29, the reputation of Lackawanna coal suffered in comparison to the more familiar coal from the Schuylkill or Lehigh anthracite regions, as even the directors admitted the "inferior character of the surface coal brought to this market in 1829." To make matters worse for the Wurts brothers, the D&H entered a feud with the Morris Canal Company, which shipped Lehigh and Schuylkill coal through New Jersey. The resulting war of pamphlets intensified the doubts raised about the ability of Lackawanna coal to serve New York markets for the future. When the D&H failed to sell twenty-four thousand tons in its second full season of operation, the supporters of the Morris Canal crowed in an open letter to John Wurts: "Your coal trade is a losing one."[27]

The winter of 1830–31 proved to be one of the worst seasons in recent memory for New Yorkers, but the arrival of severe weather was fortuitous for the D&H's anthracite business. New Yorkers suffered intensely that season. A massive snowstorm in early January left the streets covered with eighteen to twenty inches of snow for weeks. Waterways became clogged with ice, and shifting snow banks blocked the passage of both wagons and sleighs. A severe fuel shortage hit Manhattan, and the price of wood skyrocketed to over twenty dollars a cord by February—if any fuel was to be had at all. The *Commercial Advertiser* estimated that only two to three hundred cords remained in the entire city by mid-February. No relief was in sight, and between 2,600 and 2,700 New Yorkers sought refuge in the almshouse. As the cold nights continued to exact their toll on the populace, New York's ward-based charity structure organized a relief program. Fuel figured prominently in these efforts; in the Seventh Ward, for example, five hundred families received firewood purchased by public officials. Visitors appointed to check on the less fortunate brought back stories of misery that were unprecedented in contemporary memory.[28]

Where many New Yorkers saw disaster, the directors of the D&H saw an opportunity to publicize Lackawanna coal. Since Philip Hone, a former mayor of New York and a fuel philanthropist, was a manager of the D&H, the firm had a strong connection with the city's philanthropic network. Newspaper articles describing the relief efforts helped publicize the low cost and high heat of Lackawanna coal throughout the crisis. One "Committee Man" reported an encounter in which he was "much gratified . . . to find in the room of one poor woman (whose husband was in the Debtors' Jail) a good fire of Lackawanna coal made in a common fire place." "She told me it was her only fire," the visitor reported, "and that it answered all the purposes of a wood-fire for cooking, &c. and was much warmer, and did not cost half as much. I beg you to mention this in your paper, that those disposed to aid the poor with fuel, at this pinching time, may do it with less expense to themselves, than by furnishing wood." Another correspondent called "Franklin" praised the D&H's efforts to bring cheap fuel to the poor and estimated that the firm helped lower the cost of fuel by 25 percent. Not only should the D&H earn the "high satisfaction of being Public Benefactors," the report continued, "but from their own praiseworthy efforts, richly remunerate themselves in a pecuniary point of view." The Committee for the Relief of the Poor for the Fourteenth Ward helped advertise the fact that coal stove retailers such as the Salamander Works on Cannon Street offered "furnaces of clay, hooped with iron, with iron grate and chimney" for $1.50 and cast-iron furnaces of similar designs for $2.00. "It is gratifying to know that small stoves may be had for three dollars, and a barrel of the Lackawanna anthracite for 75 cents," a sympathetic reporter for the *Commercial Advertiser* wrote on February 10, 1831. "These stoves will answer for cooking. The benevolent may thus relieve the poor by a very small sacrifice of money." By February, news of the media campaign's success reached Charles Wurts in Philadelphia, where he was active in that city's fuel philanthropy. In response to letters from New York trumpeting the rise in D&H stock, Charles confidently predicted that "no doubt the consumption of coal will be more than doubled here another year."[29]

The D&H also worked through retail coal dealers, many of whom made appeals based on class. H&A Stokes and Edward Dunscomb, for example, advertised that they were delivering Lackawanna coal free of transport charge for between $6.75 and $7.00 a ton—significantly undercutting the price of dealers in Lehigh ($8.00 a ton) and Schuylkill ($9.00) anthracite. Dunscomb asserted that Lackawanna coal was "by experiment found to be the cheapest coal for cooking and also the most agreeable, not producing the head-ache as the dry heat from other coal does." He also mentioned that his coal had "the preference at the Bloomingdale Asylum, the City Hospital, and the Alms House." Although some dealers openly endorsed the coal brought to market by the D&H, others preferred to work surreptitiously with the firm in order to preserve their reputation as impartial retailers.[30]

The plan of the D&H to market Lackawanna coal to urban consumers incorporated appeals to the poor. Whether or not the directors of the D&H hoped to corner the market among lower-class consumers is uncertain, but the Wurts brothers' movement between corporate and benevolent circles provides suggestive evidence. In February 1831, Maurice Wurts wrote to Charles concerning the "great variety of cheap contrivances" that had been introduced recently in New York to burn anthracite. Maurice also mentioned the value of selling coal by the half-bushel (coal was commonly sold by the ton at the coal yard) to the working poor. By breaking up coal sales into smaller portions, they could market Lackawanna coal to a wholly different class of consumer than most coal retailers. The poor, therefore, figured prominently in the D&H's plans for future profits. "The introduction of [coal] among this class of inhabitants," Maurice claimed, "much extends it use beyond computations." Thus in New York City a corporate coal carrier very consciously tapped into philanthropic networks in order to expand markets, whereas in Philadelphia, the philanthropic institutions themselves assumed the lead in promoting fuel conversion. The Wurts brothers knew both parts by heart. By the late 1840s, the UBA was a permanent presence among Philadelphia's poor families, and Charles Wurts was its president. "The Society will have a larger supply of coal," Charles wrote in 1848, "and the poor, purchasing it, instead of procuring it for nothing, will be inspired with a disposition to exert themselves they otherwise would have not manifested."[31]

Once urban hearths adopted anthracite, either in stoves or in fireplace grates, the growth of the trade was assured. Production levels soared, and the flow of anthracite rose to impressive levels. "The increase of the Coal trade in the last fifteen years," the editors of *Niles' Register* reported in 1845, "is perhaps one of the most remarkable features of our prolific country." Anthracite colliers shipped 34,593 tons to market by various water and overland routes in 1825, according to the *Register*'s estimates. By 1834 that figure had increased to 363,861 tons; by 1845 the editors estimated that 2.71 million tons of anthracite made their way to market. This was indeed an impressive increase, and much of it was owed to individual decisions to convert hearths in urban households.[32]

Although many observers breathlessly touted the many advantages of burning anthracite to rich and poor consumers, male and female users, and cities on the whole, there were costs to displacing the firewood trade. In 1836, the humorist James C. Neal, known in some circles as the "American Dickens," wrote about the plight of a wood sawyer in the new industrial economy. His subject, an African American resident of Philadelphia named Dilly Jones, was "a man of a useful though humble vocation, and no one can saw hickory with more classic elegance." Neal reported that one night Dilly was headed home "rather late and rather swipey [i.e., drunk]" and began to ruminate on his trade. "Sawing wood's going all to smash," he lamented, "and that's where every thing goes what I speculates in." The source of Dilly's distress was clear: "This here coal is doing us up. Every since these black stones was brought to town, the wood-sawyers and pilers, and them soap-fat and hickory ashes-men, ahs been going down; and, for my part, I can't say as how I see what's to be the end of all their new-fangled contraptions. But it's always so; I'm always crawling out of the little end of the horn." In the emerging industrial economy, day laborers like wood sawyers, corders, and pilers found less and less opportunity. Even if, Dilly Jones mused, he could get work as a night watchman or catching pigs, somehow those jobs would disappear as well: "They'd soon find out to holler the hour and to ketch the thieves by steam; yes, and they'd take 'em to court by rail-road, and try 'em with biling water." Even pigs would be caught by steam and cooked immediately. "By and by folks won't be of no use at all. There won't be no people in the world but tea kettles; no mouths, but safety valves, and no talking, but blowing off steam."[33]

Neal's satirical portrait of Dilly Jones was meant to force his readers to reflect on the side effects of America's industrialization. As a human symbol of those effects, Dilly offered a romantic look back at the old firewood network that clearly was inadequate in serving urban fuel markets. And yet economic displacement would become a common setback for many Americans in the coming decades, as old trades disappeared and new industries blossomed. Although sympathetic to his subject, Neal offered that Dilly Jones's situation was hardly unique in history: "Many were the poor barbers shipwrecked by the tax upon hair-powder, and numerous were the leather breeches makers who were destroyed by the triumph of woolens."[34] Dilly probably took little comfort in the fact that charitable organizations like the Union Benevolent Association or corporations like the Delaware and Hudson wanted to make it easier for him to buy and burn anthracite to stay warm. In fact, that particular expansion of consumer choice might ring hollow for him.

But cheap mineral fuel, although the bane of wood sawyers by the 1830s, was not assured for future generations. In order for the substitution of mineral for organic fuel to continue, a massive infrastructure would need to emerge so that the flow of coal from the mining regions to cities would continue unabated. The creation of a complex industrial system that raised, accumulated, and distributed coal was beyond the capacity of a single entrepreneur, corporation, or even state. By the advent of the American Civil War, the process of raising coal to the surface, carrying it to urban markets, and then selling and distributing it to customers involved thousands of economic actors. Not surprisingly, a number of conflicts arose among representatives of the various components of the mineral fuel network. But in the end, these workers, managers, and customers all worked toward the same goal: keep the coal flowing.

3 How the Coal Trade Made Heat Cheap

By the conclusion of the American Civil War in 1865, Chicago had become one of the nation's fastest-growing industrial cities. Access to nearby reserves of bituminous coal helped fuel this growth, and three brothers, Alfred, Charles, and John Rockwell, attempted to grow a business in both mining and selling it. They owned land with rich coal seams in LaSalle, a small community about ninety-five miles southwest of their coal yard in Chicago and linked to the great city by the Illinois and Michigan Canal and the Illinois Central Railroad. Despite the rich deposits of bituminous coal and easy transportation to market, the Rockwell brothers found the coal business difficult to master. As John wrote in a letter to Alfred, "Competition in Chicago is very great and profits uncertain; it requires a large amount of money to carry on a business there and is of a speculative nature, and we cannot control it." The reason for their struggle was simple. Chicago's location on Lake Michigan allowed both anthracite from Pennsylvania and bituminous coal from Ohio to flow into the city, competing with mineral fuels from nearby sources. So much coal could be delivered that sometimes the supply exceeded Chicago's storage capacity. In 1866, for example, the firm of Kellogg & Gray rented out space in the Rockwell yard to hold five hundred tons of "Erie coal" that had arrived from Ohio. A year later, John wrote to Alfred that rival retail dealers in Chicago were reducing the price of coal shipped from Cleveland via lakes Erie and Michigan "to such figures that will crowd us hard." The Rockwells had twenty thousand tons of LaSalle coal to sell in 1866, but they faced cut-throat competition with other dealers, particularly the well-established and well-connected firm E. D. Taylor & Son, which shipped coal from nearby Will County coal mines via the Chicago and Alton Railroad. When this rival secured a reduced rate from the railroad and dropped its price for bituminous coal to six dollars a ton, John wrote to Alfred, "He can make money at that and we shall starve to death."[1]

Finding themselves squeezed by the diversity of coal streaming into the Chicago market, the Rockwells sought ways to lower prices. Writing from the mines in LaSalle, John's plan was to reduce the cost of raising coal in LaSalle to a dollar a ton in order to rival "the competition from abroad and the cheating of our friend Taylor." However, when he contacted fellow mine operators in the LaSalle region, they were "unwilling to run the risk of a strike and throw the whole trade into Taylor's hands." Labor troubles indeed followed John Rockwell's plan to cut costs, most notably when the American Miners' Association (AMA) resisted the plan to cut wages. The AMA had been a powerful trade union during the Civil War and remained a force for collective action among Illinois bituminous coal miners. When coal operators in the LaSalle region lowered the tonnage rates paid to miners in the summer of 1867, up to a thousand workers struck. Unsuccessful searches for replacement workers and coal markets in northern Illinois and Wisconsin dominated that summer for the Rockwells. "We are so mixed up with miners, soldiers, governors, and detectives," John wrote to Alfred in early August 1867, "that I can see no good ahead." The AMA and the LaSalle colliers settled by entering a cooperative agreement that linked wages and profits, while at the same time encouraging the development of a stock ownership program for workers and a cooperative store. John acknowledged that "things have been patched up again and the wheels have started once more," but the Rockwells still faced fierce competition in Chicago, where they "must sell something . . . or go out of the market altogether." Although the innovative cooperative arrangement between the miners and the colliers in LaSalle only temporarily bridged the gap between labor and capital—conflicts would resume in the coalfields after the cooperative interlude broke down—this short-lived peace avoided the heart of the problem facing coal interests: price competition. Indeed, one British trade union official who visited the LaSalle region later in 1867 denounced the miners' participation in the "violent competition" that sought to "out-sell all in the market."[2]

The struggle between the Rockwell brothers and the AMA to control the coalfields of LaSalle is a familiar story in the annals of American industrial relations. Lockouts, strikes, and work stoppages plagued the American coal-mining industry throughout late nineteenth century. No matter how hard they fought each other, however, ambitious capitalists like the Rockwells and trade unions like the AMA could not address the root cause of their conflict: the uncontrollable flow of coal to American cities and the relentless price competition that followed. Consumers and users in cities like Chicago, Cleveland, and Philadelphia doubtlessly followed these labor conflicts closely; most likely they did so by a warm stove fueled by the very same cheap coal that triggered the bitter dispute between mine operators and laborers in American coalfields. These problems certainly had roots in more widespread questions of working-class formation in the United States, in the control issues facing nearly every industrial workplace in the postbellum years, and the emergence of large and powerful corporations that defied oversight and regulation. Tectonic shifts in the political and economic landscape of the United States set the grand stage for industrial conflict that spilled into the nation's coalfields, but more immediately, the great quantity of coal reserves in the nation exerted incredible pressure on both miners and their employers by encouraging low prices and the kind of cutthroat competition that plagued the American coal trade throughout its history. A close look at the rise of cheap mineral fuel reveals two sides to affordability: low prices benefited cities but at the same time bedeviled the coal trade.

American cities across the North were never too far from productive coalfields; innovations in transportation closed any remaining gaps by the advent of the Civil War. Early anthracite canals like the Schuylkill Navigation Company and the Delaware and Hudson Canal brought cheap mineral fuel to cities along the Eastern Seaboard, facilitated the replacement of organic fuel with mineral fuel, and hastened the growth of the coal-burning stove. A generation later, railroads extended cheap coal networks to virtually every northern city (fig. 3.1). As a result, a national market for coal arose in the late nineteenth century in which anthracite from Pennsylvania mines found its way to hearths in New York City, Chicago, and even San Francisco. If the supply of anthracite dried up, bituminous fuel from coalfields in Illinois, Indiana, Ohio, and Pennsylvania easily took up the slack. By the 1870s, the United States was awash in mineral fuel as a result of this sprawling industrial system. The network that served the stove trade also created opportunities

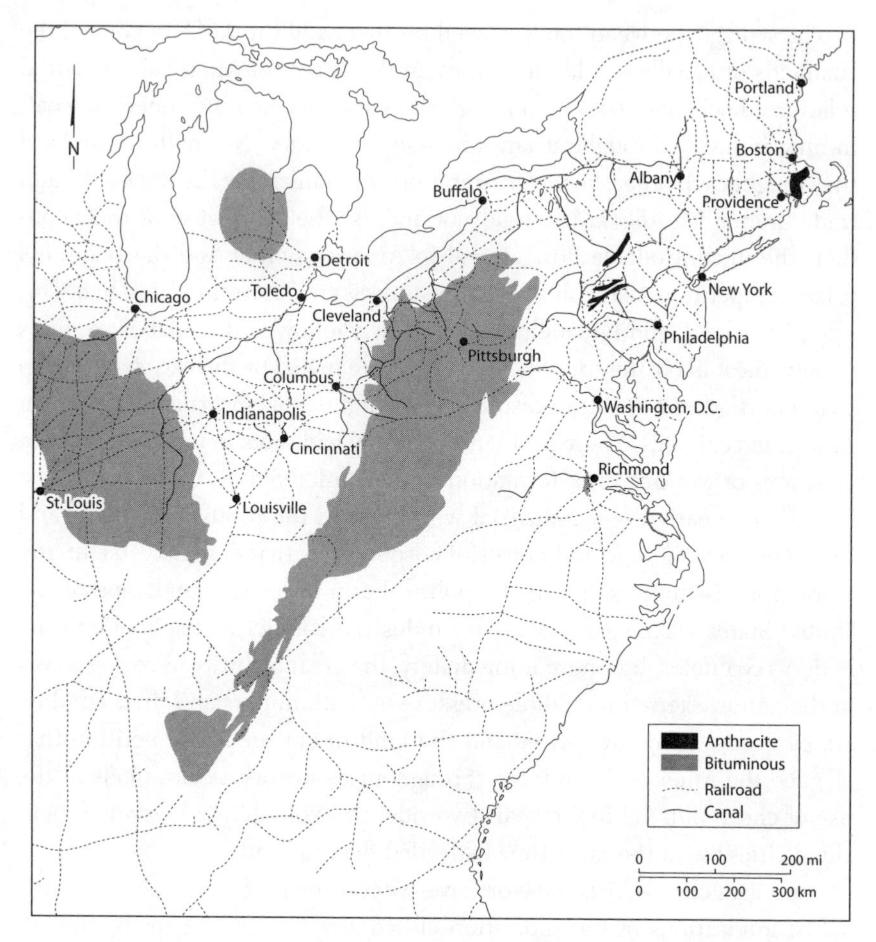

Figure 3.1. Coalfields and the US transportation grid, 1870. Bill Nelson

for colliers to reach urban markets efficiently, but not without conflicts arising. The competition between coal miners, wholesalers, and retailers benefited the consumers of mineral fuel in those cities at the same time that it vexed both capitalists and workers in the coalfields. This paradox of plenty in the coal trade helped transform the urban hearth into an industrial one in the era of the Civil War and Reconstruction.

Making Coal Abundant

"Stone coal" conquered Philadelphia and New York markets during the time that stoves replaced fireplaces as the preferred heating apparatus; eventually anthracite became a familiar commodity in markets across the American North. Philadelphia's competing network of canals, which pitted the Schuylkill Navigation Company against the Lehigh Coal Navigation Company, was the first and perhaps most successful example of cheap and effective conveyance of coal from the hinterland to urban centers. The Delaware and Hudson Canal established a stream of stone coal from Pennsylvania's northern anthracite field to New York City and invested quite a bit of time and effort promoting its product in that market. Once dealers could count on a steady supply and consumers were well versed in how to use anthracite as well as its efficiency, Pennsylvania's coalfields established themselves as the first major success in establishing an interstate coal trade. The former mayor of New York City and famous diarist Philip Hone reflected on Thanksgiving Day of 1839 on the value of anthracite, "an article indispensable for the use of all classes of our citizens" and to his city's economic health. He also highlighted the critical significance of efficient transport of "an article which, although known to exist in an inexhaustible extent in the mountains of a neighboring State, was as worthless as the soil which covered it, until the means were adopted by the construction of roads and canals to bring it to market." Hone then went on to denounce the short-sighted politicians who opposed internal improvement programs—he was, after all, a politician at heart—but his comments illustrate just what had occurred by that Thanksgiving Day. The producers and users of anthracite had created the incentive to expand the coal trade.[3]

As coal's value as a heating fuel increased, a pattern emerged in other states in which "developmental" canals provided the impetus for linking coal reserves to a growing network of fuel markets. Colliers convinced state legislators to authorize corporate endeavors that promised to exploit reserves of mineral fuel and spur economic growth or to build the canals themselves; no matter how they appeared, the new links to market aided the exploitation of the nation's massive coal reserves. Maryland's Chesapeake and Ohio Canal (C&O), for example, sought to tap into the bituminous region in the western counties of that state during the 1830s. The C&O mirrored the route of the Potomac River from Washington, DC, into the mountains of western

Maryland, passing through thick coal seams along the way. If, boosters assumed, the coal could be brought down to the Tidewater region at a low cost, then the C&O would make ample revenues on tolls, thus recovering construction costs quickly. One 1834 report in favor of the venture labeled coal "by far the most productive source of revenue on all canals where found, in Europe and America" and argued that "it follows as a matter of course that there will be no limit to the demand for exportation but the capacity of the Canal to deliver it." The opening of the Ohio and Erie Canal in 1827 unlocked the rich bituminous fields of eastern Ohio, which allowed their rapid development and, more important, a steady link to growing markets in the city of Cleveland. By the 1840s Youngstown's David Tod, who would become governor of Ohio during the Civil War, made a fortune shipping Mahoning valley coal to Cleveland, where it warmed hearths and powered steamboats of the Great Lakes. By 1848, Ohioans raised over a quarter million tons of coal, making their state the third leading producer behind Pennsylvania and Virginia. More important, as Ohio's stocks of wood fuel decreased, the residents could count on coal to replace them. Finally, the Illinois and Michigan Canal provided a developmental boost to the bituminous regions outside of Chicago when it linked Lake Michigan to the Illinois River in 1848. By 1850, the Illinois and Michigan brought 3,361 tons of coal to Chicago. Within five years, this amount had nearly doubled, to 6,701 tons. Canal shipments of coal in Illinois appeared modest in comparison with the millions of tons flowing down Pennsylvania's anthracite canals, but the advent of cheap and effective lines of transport, as Virginia's Harry Heth or the Rhode Island Coal Company would have readily averred a generation earlier, was essential to a nascent coal trade.[4]

The advent of railroad technology provided yet another way to transport coal quickly and cheaply from mining regions to urban centers. Railroad companies like Pennsylvania's Philadelphia and Reading and Maryland's Baltimore and Ohio lines competed directly with the established coal-carrying canals in those states. The result was a wider array of types of coal for urban consumers to choose from. By 1851, for example, Baltimore fuel markets consumed 163,855 tons of bituminous coal drawn from the C&O Canal and B&O Railroad; they also received 200,000 tons of Pennsylvania anthracite shipped by canal, rail, and coast. Most important, though, railroads allowed coal markets to reach across the Appalachian Mountains into the Old Northwest, linking growing urban centers there like Cincinnati, Cleveland, Louisville, and Chicago with both anthracite and bituminous fields back east. Anthracite appeared in all of those cities as the nation's rail network connected them to eastern markets. The potential for railroads to send cheap coal to virtually any location thrilled the trade's boosters. "The invention of railroads has extinguished the difficulties of transporting our coal to the remotest parts of the country where no such fuel exists," the editors of *Hunt's Merchants' Magazine* reported in 1854, "and such places otherwise uninhabitable, may be rendered cheerful and gladsome in the coldest nights of our dreary winters. . . . We could do without the gold of California, for it does not allow a single comfort to the life of man: but we could not do without our coals."[5]

By the eve of the American Civil War, urban residents across the nation could count on supplies of coal from a myriad of sources. Historian Christopher Jones estimates that by 1860, 90 percent of Philadelphia's and 85 percent of New York City's home heating demands were met by anthracite. This increased flow of coal into American cities reordered traditional labor patterns, created new cities where none had existed, and reshaped the urban landscape. Most important, though, it finally ended the problems of intermittent supply that plagued cities of the Early Republic and threatened to put a damper on urban growth in the Old Northwest. Unlike Philadelphians and New Yorkers in the 1810s, Clevelanders and Chicagoans in the 1850s did not have to worry about disappearing stocks of wood; nor did they fret over the fuel efficiency of open fireplaces and leaky chimneys. Armed with stoves and coal, Americans across the urban North conquered winter's bite by constructing a massive industrial system that delivered heating fuel from the depths of the earth to the doorstep of the consumer. Like a river that winds its way through the countryside, coal flowed from sparsely populated mining regions into the burgeoning cities of the North.[6]

At the headwaters of this river lay the basic industrial unit used to extract coal, known as the "colliery." Collieries encompassed the wide range of above- and belowground structures—mine shafts, winches for lifting cars and elevators, toolsheds, steam engines, breakers, railroad or canal connections—each of which had a particular function in the process of extracting coal from the ground. Some colliers were individual proprietors who leased their land and worked the seams themselves; others worked as salaried managers for large coal companies. A collier's work varied greatly across the diverse coal regions of the United States. In the anthracite and bituminous fields of Pennsylvania and Maryland, the formation of mountains tilted the coal seams at various angles. In some cases, the seam broke the surface, and colliers followed them into the ground with a "drift" or "slope" mine that burrowed into the side of the hill. In flatter coal regions, such as the bituminous fields in Illinois, the drift and slope mines found the seams at riverbanks. Many seams in the American Midwest were easily accessible at first, and miners could dig the coal without an elaborate system of tunnels and shafts. When these easily mined seams were exhausted, colliers constructed shaft mines, which burrowed vertically into the ground, and they often intersected the coal seam at a right angle.[7]

Collieries in the nineteenth-century United States sunk shafts hundreds of feet deep, creating a host of engineering challenges. Keeping the ceiling of the shaft intact, ventilating the work areas, and keeping them relatively dry took copious amounts of time and energy. Deeper mines saw water seep through the walls of the shaft, and shallow ones collected rainwater; no matter the source of moisture, the working areas of a mine were nearly always wet and the air damp. American collieries employed steam-driven pumps in the Sisyphean task of keeping mines dry, but they could not remove all the water. Methane gas, which miners colloquially called "firedamp," also permeated many coal seams, creating a constant fire hazard. The early practice of placing furnaces near the opening of mine shafts to draw fresh air from the surface meant that ventilated air came at a great risk to life and limb. Eventually, colliers resorted to fans to recycle fresh air. Yet these innovations never seemed able to keep up with the deleterious effects of working underground, and they made the seemingly straightforward task of coal extraction a complex and dangerous task for American collieries. "A coal mine must present an unusually exciting scene," Edmund Ruffin wrote in 1837 after visiting a Virginia colliery, "there being added to the usual matters of interest, the increased apprehension of danger—the rapidity of the excavation, by the concentration of numbers in a small space—and successive abandonment of each portion—and by the continual sounds of cracking and crushing timbers, and falling in of masses of the stony roof, in places not many yards distant, and which were excavated by the miners but a few days, or perhaps but a few hours before."[8]

Simply tearing coal from a seam and raising it to the surface is a difficult job. Until the advent of cutting machinery in the late nineteenth and early twentieth centuries, most American coal was mined using techniques that had been brought over from Europe and adapted to the North American coalfields. Individual miners operated, for the most part, as independent contractors with either proprietorships or coal-mining companies, who paid a piece rate, usually per ton, for the amount of coal raised. Throughout the nineteenth century, *miner* referred to the skilled workers in charge of underground work—this did not include the laborers who shoveled and transported the coal to the surface. Each morning the miner and his crew of one or two laborers walked or crawled to their individual chamber within the mine, where the first order of business was to assess the best way to loosen the coal. Sometimes, a miner could undercut the seam, working flat on his back with a pick and shovel and propping the overhang with "sprags" to create a space at the bottom of the face. Once the seam was undercut, miners used wedges to shear large chunks of coal from the seam (fig. 3.2). If the coal proved too hard, miners could bore a hole into the face with a hand-operated auger, pack it with a small amount of black powder, and detonate it. Blasting sometimes shattered the coal into less desirable small shards, but working with stone-hard anthracite in Pennsylvania often left miners with little choice. Once the coal was on the floor of the mine, laborers shoveled it into small cars for transport to the surface. If a mine used the common "breast and pillar" method, the miners left some coal standing as a roof support, the pillar, while they carved out a major portion of the seam, the breast, where they extracted the coal. Or the miner could set his crew to propping up logs in the place of the coal pillars. Since miners were not paid for this "dead work" (work not directly involved in extracting coal) in the mines, they tried to accomplish it as quickly as possible. By the time the coal was cleared, the miner would be on to the next area of the face to plan the next round of undercutting.[9]

This was how coal mining was done in theory. In practice, American coal miners faced one of the most dangerous and unpredictable workplaces in the industrial world. Across the nation, miners faced a host of daily challenges to their survival. A shaft collapsing, an explosion of the methane gas that seeped from many coal seams, or a fire breaking out in the mine's cramped spaces could be deadly. Americans working in coal mines faced much more mundane miseries as well. Mules used for hauling coal and other equipment usually stayed underground for much of the year; their feed and droppings lured mice and rats. Well-constructed latrines or sewage systems were rare, so miners and animals alike added to the foul quality of the air, which the imperfect ventilation systems struggled to abate. Pumps could not remove all the water from mines, so workers often walked or crawled to the face of the seam through a foul slurry. Creaks and cracking noises, explosions distant and near, and constant clatter of shovels and picks made for a noisy workplace.

Figure 3.2. A coal miner at work. Coal mining was difficult, dangerous, and unpredictable. Few artists could capture the severe working conditions, but this print demonstrates the difficult task of undercutting the coal seam by hand. *Harper's New Monthly Magazine* 29 (1864): 163

•

Coal mines looked clean and logical on the drafting tables of mining engineers on the surface; down at the seam they were noisy, pungent, and cramped workplaces that demanded hard and monotonous labor often punctuated by terrifying moments of potential disaster.[10]

Once the miner had "won" the coal from the seam and it had been transported to the surface, the aboveground labor force went to work processing it for transport. Collieries had many buildings at the surface of the mine—winches for pulling cars or elevators, machine shops, equipment sheds, and changing rooms and bathhouses for workers. The most important structures at an anthracite complex were the "breakers." These massive hundred-foot high buildings stuck out in what were otherwise barren landscapes in the coal regions of America. The coal started its journey through the breaker at the top, where it would be dumped into a hopper that shunted it into a number of screens designed to remove slate, mud, and other materials from the raw coal. Once removed from the coal, the waste material, or "culm," did not usually travel far from the mouth of the mine; workers deposited it in a "culm pile." The coal continued to work its way through the breaker, aided by gravity and the "breaker boys"—or children too young for underground work—who picked any remaining slate or rock from the coal as it whizzed underneath them on a conveyor belt toward a series of crusher rolls that broke it up into various sizes. Then the coal traveled over a series of metal screens with different-sized gaps that automatically sorted the coal into marketable categories such as "lump," "egg," "chestnut," and "pea." The final task for workers at the colliery was loading the coal into canal barges or railroad cars through a structure called a "tipple." Larger breakers might have an automatic sorting system that distributed the various sizes into different containers; smaller collieries hired workers to shovel the coal into waiting boats or cars (fig. 3.3). Bituminous mines, where breaking the coal into different sizes was less critical, often sorted impurities and loaded cars directly at the tipple. Regardless of the size or complexity of the colliery, by the time the coal was ready for its trip to market, it had exacted a high cost in life, limb, and labor.[11]

After it had been raised and sorted at the tipple, colliers transported the coal to urban centers as efficiently as possible. Canal companies that owned or leased coal lands, like New York's Delaware and Hudson, shipped it to market via "captive" canals and railways that they owned or controlled themselves. In cases where a variety of independent collieries were at work, such as Pennsylvania's Schuylkill County, the coal accumulated at a regional center like Pottsville, described by *Hunt's Merchants' Magazine* in 1846 as the "grand depot" of coal "as well as the place of shipment" from a spiderweb network of short, lateral railroads connecting various collieries. Most rail carriers operated on razor-thin margins and adjusted toll rates to encourage heavy traffic. "The profits on coal are so small," engineer Herman Haupt reported to the Pennsylvania Railroad in 1857, "that a slight reduction just sufficient to give a profit to the operator may quadruple the shipments, and increase tenfold the net profits of the transporter." Cities along natural waterways benefited from the cheap transport of coal in steamers or sail-powered schooners. Western cities, like Cincinnati and Louisville, received coal shipped cheaply along the Ohio River, while cities along the Great Lakes, like Cleveland or Chicago, tapped into a thriving lake traffic. At first, cities in the West imported eastern anthracite or bituminous coal at a relatively expensive rate, but as the nation's rail network expanded, the availability of both local and distant coal increased. For example, the Illinois Central reported in 1856 that coal dealers in Cleveland dominated Chicago's coal supply, despite being hundreds of miles away. Once the Illinois Central's "facilities for production and distribution are fully systematized," J. W. Foster predicted, "cheap fuel will be at the

Figure 3.3. An early coal breaker in Pennsylvania. Once brought to the surface, coal went through initial processing before shipment. After it rose to the top of the structure via a conveyer belt and was shattered by breakers, it traveled down again through the tubular screens depicted in the center of the image, where it was automatically sorted into the waiting carts below. Once sorted into different sizes such as "egg," "nut," and "pea," the coal could be shipped to coal dealers, who then priced it by both size and quality. *Illustrated News* (New York), 15 Jan. 1853

command of every one; and undoubtedly each year will reveal new sources of supply." Across the nation, the predictions of the Illinois Central proved accurate; the flow of coal increased unabated in the 1840s and 1850s.[12]

When coal reached its urban markets, it encountered an elaborate distribution system that far outpaced the traditional wharfs where farmers unloaded cords of firewood for inspectors in the cities of the Early Republic. In their place, huge coal yards sprung up—almost overnight—around wharfs and railroad terminals across the nation, drawing upon a small army of workers to keep the coal network humming. Philadelphia saw 1,339 vessels carry coal to the city by 1833—a trade that the editors of the *Commercial Herald* observed had "been literally created within the last six years." On a single day in 1835, thirty-five barges laden with 1,651 tons of coal arrived at the Schuylkill wharves. Coal heavers unloaded this cargo, hauled it to coal yards, and then dealers employed more manual laborers to deliver it to their customers. Beyond the coal yards, the distribution of coal varied by consumer. Coal dealers sold by the ton directly to affluent customers, who could afford to lay up a supply of coal for an entire winter's season. Less affluent consumers purchased their coal indirectly from smaller retail outlets such as grocery stores. In 1815, a New York City directory listed seven merchants who sold coal but only a single coal yard on Warren Street near the Hudson River. By the eve of the D&H's entry into the New York market in 1830, five coal yards dotted both the Hudson and East Rivers, and the city listed a coal inspector, a full-time coal carter, and eight merchants who sold coal. In 1845, eight coal yards, three inspectors, and at least eighty-six coal merchants were spread throughout the city. Competition was fierce among coal retailers such as small-scale grocers and dry-goods merchants; they all sought to secure a supply of coal as cheaply as they could so that retailing it to their customers could earn at least some profit. Francis Doremus, a grocer on Manhattan's Pearl Street, purchased tons of coal from 1834 to 1838 from five separate coal dealers who operated out of coal yards on both the Hudson and the East River—hardly a large distance. Doremus's willingness to buy from different suppliers suggests that price rather than loyalty or familiarity drove most wholesale purchases.[13]

As coal passed from buyer to buyer, the retail price increased. Even though colliers and dealers constantly complained about prices being too low, poor consumers saw a larger portion of their income going toward heating fuel. From their perspective, coal was neither cheap nor easily obtained. When the journalist George Foster visited a "low and mean" store in New York City's notorious Five Points neighborhood in 1850, he noticed a row of small boxes from which customers could purchase nails, tobacco, charcoal, anthracite, and other items in any quantity, even a single cent's worth. On one shelf he saw a pile of firewood priced at "seven sticks for a sixpence, or a cent a piece." Foster highlighted the wide array of goods as well as the huge profit margins—sometimes in excess of 500 percent—that Mr. Crown, the proprietor, made selling commodities in such small quantities to desperate consumers. "His customers, living literally from hand to mouth, buy the food they eat and even the fire and whiskey that warms them," Foster lamented, "not only from day to day, but literally from hour to hour." This was yet another example of the two-sided nature of affordability. As coal became more ubiquitous in urban hearths, households became more dependent upon it for warmth. But being at the tail end of the many transactions that brought the coal to urban retailers raised its price to levels that the most vulnerable Americans could barely afford. Urban families living on the margins of the economy scrounged for dropped coal along rail lines, picked lumps out of trash bins, or burned discarded wood. For an unemployed laborer or a single working mother, the low wholesale price of coal did not mean cheap fuel.[14]

Competitive forces drove wages down all along the coal distribution network, which caused further problems for working families and often threatened to cease the flow of mineral fuel into cities. In May 1835, the Schuylkill coal wharves fell silent as, in the words of Philadelphia's *American Daily Advertiser*, "the coal-heavers . . . determined that ten hours work per day is as much as comports with their comfort and dignity, and . . . therefore decreed that none shall work longer." About three hundred men banded together to prevent seventy-five barges on the Schuylkill from landing and discharging their cargo, led by a sword-waving man who "threatened every man with death who dare[d] lift a piece of coal." In June 1835, boatmen on the Schuylkill coal barges joined the coal heavers and also struck to demand higher wages. Even Philadelphia's wood sawyers, "who [did] not consider themselves . . . too humble to have a voice in the matter," called for an increase in their fee for cutting wood to 50 cents per cord (up from 40) for oak and 75 cents (up from 60) for hickory. That July the boatmen working the Schuylkill Canal struck for higher tolls, arguing, "As long as we leave our prices to be related by any persons interested in the coal trade, we will always be kept as slaves." These strikes ended peacefully, but a year later the coal heavers struck again for a twenty-five-cent increase and enlisted Philadelphia's General Trades Union in their cause. This time Philadelphia officials threw the strikers in jail, set bail at $2,500, and ultimately broke their walkout. Formal organization of the coal heavers withered in the economic crisis following the Panic of 1837, but the threat of an impromptu strike by either the shoreside workers or barge tenders still loomed large and had the potential to disrupt the supply of coal at any time. Even if the system ran smoothly, the sheer amount of coal could clog distribution networks. In 1846, John Dupuy estimated that a wagon laden with coal could take a full day to travel the length of Philadelphia's Broad Street.[15]

In western cities, coal distribution networks had taken a different shape by the advent of the Civil War. Early residents of Chicago, for example, burned abundant local stocks of wood. When the first shipment of coal aboard the

General Harrison arrived at Chicago's lakefront wharfs in 1841, the firm Newberry & Dole had a difficult time selling it. By the 1850s, however, the story was quite different, as the city's explosive growth and its transformation into a major center of agricultural agglomeration and industrial production placed new pressures on Chicago's fuel markets. The city's extensive rail connections allowed Chicagoans to draw upon distant markets for commodities. Chicago, then, drew first from national markets for coal before it relied upon local supplies—an inversion of the development of the fuel distribution networks of the East. When Arthur Meeker set up his coal dock and yard on North Market Street in 1857, he drew upon his connections and experience with Pennsylvania's Lehigh Valley Railroad Company to import coal from the East. In that year, 134,043 tons of coal arrived in Chicago via Lake Michigan, as opposed to 30,671 by railroad. In 1858, the LaSalle Basin's Little Rock Mining Company asserted that once coal could be shipped efficiently along the Illinois Central's rails, Chicago consumers would see savings of over 60 percent on the price of anthracite and 33 percent on bituminous coal. By 1860, colliers in Illinois raised an estimated 857,600 tons of bituminous coal—the vast majority of which ended up in Chicago's furnaces and hearths.[16]

The Civil War Transforms the Coal Trade

By the time of the Civil War, then, an impressive network of coal flowed across the cities of the North, as the trade benefited from the region's dense concentration of ports, canals, and railways. Although mineral fuel figured prominently in manufacturing and transportation, the domestic consumption of coal rose to unparalleled levels by the 1860s. A quick snapshot of the trade from a national perspective suggests that both a dramatic increase in production and a gradual decline in price encouraged fuel consumers to purchase more and more coal to serve their home heating needs. From 1850 to 1860, for example, the estimated value of anthracite production more than doubled, from $5.3 million to $11.9 million, while the estimated tonnage increased from roughly 8.4 million to 20 million. Prices varied by location and by the type of coal, but a 1907 estimate by the American Iron and Steel Association found that the average price of a ton of anthracite fell 6.6 percent during the 1850s, from $3.64 in 1850 to $3.40 by 1861. This is, of course, a generalization, as prices fluctuated by the month, depending on weather or strike-related disruptions in supply or general inflationary trends in the economy. Generally, however, the price of a ton of anthracite ranged from three to five dollars per ton from the 1840s until the outbreak of the Civil War in 1861, with the general trend being downward. In bituminous coal regions, the trends were a bit more volatile, as new sources contributed to national markets at an uneven pace. But even among producers of soft coal, as the Rockwell brothers would soon discover, falling prices combined with increased production helped consumers at the same time that they placed great strains on colliers and miners.[17]

The outbreak of the Civil War provided both short- and long-term challenges for the American coal trade; these developments would have major repercussions for mineral fuel consumers in the postwar period. The first and most immediate change was an upsurge in prices. As both state and federal officials ratcheted up their efforts to raise armies, the business prospects for American colliers surged. Prices shot up during the war years, as purchasers anticipated an increase in the demand for mineral fuel at the same time that disruptions in supply could be both sudden and devastating. The invasion of Pennsylvania by the Army of Northern Virginia in 1863, for example, pushed prices skyward, as colliers in anthracite country were convinced that their region was the target of the Confederate operation. As it turned out, General Robert E. Lee had little interest in occupying the stone coal mines, but his Gettysburg campaign did offer a good example of how the nation's coal trade could be disrupted quite suddenly or dealers could take advantage of wartime confusion to gouge the market. "It is needless to state that that rebel army is not in need of coal, and we are assured that not a miner in Pennsylvania has volunteered to defend the State," the editors of the *Chicago Tribune* caustically commented in the summer of 1863. "Consequently there is no cause for the advance in prices." For whatever reason, though, coal underwent a massive price spike. On average, the American Iron and Steel Association estimated that wartime demand pushed anthracite prices up by 147 percent in 1864 and pushed bituminous coal prices up by 120 percent the following year. In certain locales, the price spike was even more pronounced. The price of Schuylkill white ash lump coal in Philadelphia, for example, which was listed at $3.28 per ton in the early winter of 1860, shot to an all-time high of $10.75 per ton by the summer of 1864. Cincinnatians in the winter of 1864 paid between twenty and twenty-five cents per bushel for bituminous coal shipped on the Ohio River—a fourfold increase from the range of five to seven cents a bushel in the winter of 1860–61. The *Chicago Tribune* reported anthracite prices to be between $22 and $24 per ton by late 1865. Although a healthy portion of these high prices could be attributed to the general inflationary trends that hit the northern economy during the Civil War, the sharp break in the downward price trend convinced many colliers that more money was to be made from their trade after 1861.[18]

The experience of the Boston merchant Supply Clap Thwing offers an example of how this wartime spike in prices shook up the long-term antebellum trajectory of the coal trade. By the time Thwing broke into the coal business in the 1840s, he had become a global trader dealing with a wide variety of goods, with business contacts in New Orleans, Philadelphia, Liverpool, Havana, and Calcutta. To an associate in New Orleans, he wrote in 1847 that he was "exceedingly interested" in the coal trade, which he saw as "a beautiful business, involving no risk." Only two years later, however, Thwing noted that "competition is rife and our success uncertain," as he grappled with various sources of coal coming into Boston via water and rail connections. For the next decade, he struggled to earn money in a trade characterized by intense price competition, Boston dealers willing to play one supplier off against another, and colliers shifting traffic to the lowest bidder. "I think I shall resign my coal agency and try to get some business from New Orleans," a despondent Thwing wrote in 1857. Thwing stuck with coal but complained in 1859 that "the dealers here have no union among themselves & if a price is established among them very few will stick with it—and they are always trying to undersell each other." Most antebellum coal merchants shared this frustration; as the national supply of coal increased by volume and location, profit margins narrowed. By 1862, however, Thwing's tune had changed. Not only was he making money in Boston—he reported that anthracite prices were rising to nine dollars a ton—but Thwing looked to expand his business along the East Coast and beyond. That April, he shipped two hundred tons to San Francisco in order to cultivate markets on the West Coast. His invoice to the dealer in California broke down to a rate of $18.49 per ton. Hundreds of merchants, dealers, and jobbers shared in Supply Clap Thwing's sudden burst of entrepreneurial optimism; wartime price spikes meant that easy money was to be had in the American coal trade.[19]

The Civil War price spike revolutionized the structure of the American coal trade. Take the way in which coal-mining firms were organized, for example. Most antebellum colliers were independent proprietors with small landholdings and a handful of employees; major corporate colliers like the

Lehigh Coal and Navigation Company were relatively rare. High prices, though, spurred some colliers to seek an expansion of their operations. Acquiring more mineral lands, expanding shafts, purchasing machinery, and hiring more miners required vast amounts of investment capital, so securing a corporate charter from the state legislature became a pressing issue for many colliers. The problem was that established mining interests in large coal-mining states like Pennsylvania limited the number of corporate mining firms during the antebellum years. A Pennsylvania senate committee found in 1834 that in mining and manufacturing, "every charter or act of incorporation, is to a greater or lesser extent an infringement upon the natural rights and liberties of the people—and their natural tendency is to monopoly." Politicians from the areas dominated by proprietary—that is, noncorporate—mining firms thus viewed corporate mining as anathema to the spirit of fair competition and fought hard to keep corporations out of their regions. If charters appeared at all, they were likely to be created under the process of "general" incorporation. This administrative act allowed firms to secure a corporate charter without legislative approval, thus bypassing the logrolling, vote-swapping, and out-and-out bribery that often accompanied the granting of special charters. In return for making incorporation an administrative process outside of political control, general chartering laws put an upper limit on the capital that a firm could raise, restricted the operations it could engage in, and created a boilerplate organizational structure. The sum of all this meant that general charters created companies with limited power and influence. By the outbreak of war in 1861, thirteen states had made general incorporation mandatory. Even in states that still allowed special charters, policy makers used general incorporation to target certain industries or regions for development, thus exerting control over the pace of chartering and its impact on individual proprietorship. In Pennsylvania's coal trade, for example, legislators passed general incorporation acts in 1849 and 1854 that excluded incorporation in coal-producing counties with established proprietary interests and introduced incentives, such as the ability to construct railroad lines, for firms seeking incorporation in less developed coal regions. So whether a state had a special or a general chartering regime or some combination of both, the general view was that corporations required oversight and strict rules guiding their creation.[20]

The economic boom provoked by the Civil War encouraged state legislatures to increase the pace of creating mining corporations precisely when more and more colliers sought charters. One journalist, C. B. Conant, noted the heightened wartime demand for coal in 1865 and found it "in the interest of capitalists to encourage this unreasoning outcry, because they take advantage of it." "Capital has no bowels or patriotism," Conant argued, "and capitalists are instinctive." Colliers and capitalists—these two groups did not always overlap—sought corporate charters at an unprecedented rate during the Civil War. In Ohio, mining corporations accounted for 36 of 91 (39.5%) charters that were issued in 1864 and 245 of 358 (68.4%) issued in 1865. Illinois issued twenty-five charters for mining companies in 1865, which made up 14 percent of the corporations created in that year. As the nation's leading coal-mining state, Pennsylvania experienced perhaps the most dramatic spike for both general and special incorporations in coal mining. Strong anti-chartering interests centered in the southern anthracite region's Schuylkill County had led the antebellum effort to keep corporations out of the state's anthracite fields. Thus the general chartering laws of 1849 and 1854 applied principally to the state's western bituminous fields. All this changed during the Civil War, when the urgent need to mine coal outweighed the ideological fretting about corporate mining. The Pennsylvania legislature passed a liberal general chartering law in 1863, and legislators issued special charters at an unprecedented rate. Politicians at the forefront of this movement described the older, informal restrictions on chartering as a "narrow, illiberal flood-tide of prejudice cherished throughout Pennsylvania" championed by "every demagogic political movement" that had held wealth "in leading strings of ignorance and bigotry." The Civil War saw such long-standing practices in the coal trade overthrown nearly overnight.[21]

A different sort of Civil War battle raged in the legislative corridors of Harrisburg, Pennsylvania, in 1864 when the supporters of the massive Mammoth Improvement Company sought to push through their incorporation bill over the objections of anticorporate forces. The creation of such a coal-mining corporation that would be authorized to hold thousands of acres of land and would be capitalized at $1 million would have been untenable before the war. For years, the Pennsylvania legislature had restricted the number of corporations operating in areas dominated by individual proprietorships. So the fact that the Mammoth charter would create a company in the heart of the Pennsylvania coal trade's anticharter country dominated by small proprietors, Schuylkill County, was even more astonishing. "It is an outrage upon the rights of Individual Operators, an outrage upon the business community and upon the people—in fact it is *rank treason to the County*," the region's *Miner's Journal* thundered, "and our members who participated in the outrage by not resolutely opposing this bill ought to be indicted as conspirators to destroy the County, and hand all the people over to the tender mercies of Mammoth Corporations." When the charter passed, albeit under the less menacing name of the "Preston Improvement Company," observers blamed the unique conditions of the war for the sudden change in policy. "The high price of coal and iron has created a furor among the capitalists amounting almost to a mania, and the files of both houses are filled with bills for chartering new Coal and Iron Companies, and supplements to those already in existence," an exasperated columnist for the *Miner's Journal* reported only a few days later. "These companies are forming in all sections of the State, and the easy manner in which they pass proves that opposition to them would be futile." In 1863, only one mining corporation was to be found in anthracite-heavy Schuylkill County, and it accounted for a miniscule amount of the region's production. A year later, twenty-five new corporate mining firms were formed. By 1865 that figure had reached fifty-two, and they produced about half of the county's coal. Pennsylvania mining corporations founded in 1864 alone accounted for over a third of both the number of existing firms and the authorized capital in the state. The exigencies of the Civil War removed the political barriers to corporate coal mining in Pennsylvania. Other states followed, and it became clear that once the war concluded, the American coal trade would be dominated by large, well-financed corporations.[22]

At the same time that large mining corporations emerged to dominate the nation's largest coal-producing state, Pennsylvania's miners found themselves increasingly without power to affect wages and working conditions. When strikes broke out in Pennsylvania's anthracite regions in 1862 and 1863, state officials sided with corporate interests, effectively using the war to undermine labor organization in the coalfields. Mine operators, many of them newly organized under Pennsylvania's liberal wartime chartering regime, cited violent resistance to the military draft established by the Conscription Act of 1863 to demand that state and federal officials quell the rising power of the miners. Federal provost marshals, together with regular army troops, imposed martial law in mining regions, ostensibly in order to suppress draft rioters. In fact, the presence of federal troops broke the back of the burgeoning miners' unions. This pattern continued after the war as well, state officials often citing wartime precedents to justify their support for large corporations at the expense of workers' organizations. Mine operators sponsored the Coal and Iron Police Act in 1866, for example, which created a quasi-public enforcement agency designed to break strikes and repress labor organization under the pliant guidelines of "protection of private property" and "safeguarding the interests" of the public. And just as draft resistance provided the wedge for undermining collective action by workers, the Coal and Iron Police protected property in ways that frustrated attempts to organize workers. Even worker-friendly acts seemed to favor the coal companies after the war. In 1868 the legislature passed an eight-hour workday but included a loophole that excluded contracted workers—the vast majority of mine laborers—from the provisions of the act.[23]

The trends toward organization that characterized the Pennsylvania trade were repeated throughout the nation's coalfields. In Ohio and Illinois, for example, bituminous coal miners sought wage increases and control over working conditions both through spontaneous resistance and by forming trade unions. Localized strikes broke out throughout the Civil War, but miners also sought long-standing changes in their relationship with colliers. The American Miners' Association emerged from the Belleville bituminous field outside of Saint Louis in 1861, in part as an effort to stabilize wages by raising prices. Its inspirational leader, an English emigrant miner named Daniel Weaver, made an eloquent plea for collective action. "The insatiable maw of Capital would devour every vestige of Labor's rights," he argued, "but we must demand legislative protection; and to accomplish this we must organize." Weaver believed that the AMA must "steadily discountenance" strikes and argued in 1861 that "capital, as well as labor, must have its due." The unsettling impact of wartime shortages of labor, coupled with price spikes, forced the leaders of the AMA to abandon their gradualist philosophy. When a series of strikes rocked both the Belleville and LaSalle mining districts in early 1863, AMA officials helped coordinate action and provided financial relief for the strikers across the state of Illinois. Using its new periodical, the *Weekly Miner*, to coordinate action, the AMA expanded into Ohio and western Pennsylvania and threatened to emerge as a national labor organization to combat the rising power of mining and railroad corporations.[24]

Mine operators in Illinois, however, provided their own collective response to the AMA's rising power in 1863, when representatives of twenty-four corporate and proprietary collieries convened in Chicago to denounce the "secret society formed by the miners" to increase the price of coal in Illinois. The signees of the convention agreed that they would not deal with the AMA but instead would "hire and discharge persons as the exigencies of the business and the conduct of those individuals may compel them to do." At the same time, the Illinois legislature passed the "LaSalle Black Laws" by a bipartisan vote in 1863, the name being an overt reference to the successful strikes in northern Illinois earlier that year. These laws prohibited the entering of a coal mine in order to induce others to leave it and made it illegal for two or more individuals to deprive a landowner of "its lawful use" through "suggestions of danger or other means." As a way to limit strikes in the region and frustrate the growth of labor organizations like the AMA, these "Black Laws" also set up stiff fines for the violation of these rules. As in Pennsylvania, colliers in Illinois marshaled the power of the state to assert their control over the trade at the expense of workers' collective action. The AMA did not survive long in the postwar years, and a truly national union for coal miners would not reappear until the formation of the United Mine Workers of America in 1890.[25]

At the point of production, high coal prices and the allure of larger profit margins during the Civil War had a profound effect upon coalfields by spurring colliers to create more corporate mining firms, consolidate their control over labor in the mines, and attempt to remake a trade that previously had been defined by razor-thin profit margins and cutthroat competition. Miners saw the opportunity to organize and demand higher wages and to use trade unionism to counterbalance the power of corporate and proprietary colliers. America's coalfields had been transformed in a lasting fashion during a relatively short period of time, and the punch-counterpunch between mine operators and miners would continue in various institutional and organizational forms over the next century. Railroads, for example, increased their participation in the national coal trade. Across the North, railroad companies benefited from the unique economic conditions created by the Civil War and emerged from the conflict in strong financial shape; in coal-producing states like Pennsylvania the loosening in state oversight of corporate entities offered railroads a golden opportunity to consolidate their gains on coal carrying. As they came under less scrutiny from state policy makers, railroads often touted the liberal provisions of their charters to investors. "It is believed that no charter more favorable for stockholders was ever passed by the Pennsylvania Legislature than that of the Summit Branch Railroad Company," one 1863 prospectus bragged. To drive this point home, the Summit Branch Railroad Company not only publicized its large authorized capital and generous provisions but also provided investors with a summary of the antebellum restrictions placed upon other railroads operating in Pennsylvania.[26]

Economic conditions during the Civil War encouraged the formation of "coal exchanges" in American cities, as wholesale and retail coal dealers attempted to control prices and regulate their trade. By circulating information about production levels, streamlining transport and distribution, and attempting to stabilize the price of coal, these exchanges sought to impose order on the open market in mineral fuel. The Philadelphia Coal Exchange organized in 1863, the same year in which strikes and draft riots in anthracite country rocked the trade. Some conservative elements viewed the exchanges as a counterbalance to the growing power of unions in Pennsylvania's coalfields. One antiunion commentator in Philadelphia's *North American and United States Gazette* demanded that the Philadelphia Coal Exchange "sustain the honest laborer and adopt such a policy as would counteract the influence of wicked conspirators." To demonstrate its patriotism, the Philadelphia Coal Exchange donated large amounts of coal to the city's "sanitary fairs," which were civilian-organized fund-raisers for Union soldiers. In 1864, it organized and funded its own military regiment, the 197th Pennsylvania Infantry, commonly known as the "Coal Exchange Regiment." This short-lived unit guarded Confederate prisoners at Rock Island, Illinois, and was disbanded by the end of the year; nonetheless its value as propaganda among coal consumers seemed to justify the expenditure. While miners conspired to raise the price of coal and resist the draft, the Philadelphia Coal Exchange argued, the dealers in coal fought to lower prices and put soldiers in the field.[27]

Coal exchanges also scapegoated miners directly as the cause of high wartime prices. The Pittsburgh Coal Exchange predated the Civil War but expanded operations during the conflict in response to a disorganized transport system from local mines to that city's markets. In an attempt to halt what it considered predatory rates on shipping, the exchange attempted to fuse the interests of mine owners, dealers, and consumers into a collective response that set rates for both mining and transporting coal. Failing to do so, the Pittsburgh Coal Exchange adopted a more stridently antilabor stance. In 1864, the exchange denounced a successful strike that raised the pay for local coal miners from four to five cents per bushel and resolved "to get a unanimous decision to resist the demands of coal diggers." Miners responded that the high price of coal was not their fault but that of "the rebellion, and the avaricious disposition of the bosses."[28]

Conditions during the Civil War helped usher in a national coal industry in the United States by accelerating the integration of corporate mining interests and railroads into the distribution network while blunting the impact that organized labor could exert on production. The appearance of coal exchanges across the industrial North in the postwar period is evidence of this important change and of the desire among many actors along the mineral fuel network to harness it. Boston, Cleveland, Chicago, and Cincinnati all had their own coal exchanges by the end of the 1860s. These organizations did collect and disseminate information about production levels, provide quality control over weights and standards, and offer a collective voice for dealers in coal, but they never succeeded in regulating the flow of coal into cities as much as they hoped. Instead, the coal exchanges followed the pattern established by corporate mining and the growth of railroads; larger entities cast a wider shadow over the national coal trade, but none of these players were large enough to dominate it. The overlapping transportation and distribution networks that distributed coal to cities grew in complexity and scale as a result of the entrepreneurial initiatives of individual and corporate actors and the political policies that allowed easy chartering created by Civil War–era state legislatures. Following the wartime spike in prices, the nation's mineral fuel networks exacerbated the tendency toward cutthroat competition. Nature had always made coal abundant in the United States; the politics of this particular time in American history helped make it cheap for generations.

Reconstructing the Markets for Home Heating Fuel

Of course, urban households that purchased coal suffered from high wartime prices at the precise moment that colliers, carriers, miners, and dealers sought to profit from them. Critiques from both public and private sources highlighted the problems inherent in the consolidation of coal-carrying entities and targeted areas in which the flow of coal was intentionally bottlenecked by corporate interests to keep prices high. In New York, for example, the state senate commissioned a study led by the former state deputy engineer and surveyor for New York Samuel Sweet to investigate the high prices of coal. In his 1865 report, Sweet concluded that the choke point in getting anthracite to New York's consumers was Pennsylvania's lack of oversight of its network of coal canals and railroads. Sweet concluded that when Pennsylvania sold off the last of its publicly owned canals in 1858, she "not only placed herself at the mercy and interests of private corporations, but this act imposed heavy burdens upon the coal consumers of this and adjoining States." Sweet provided a breakdown of the actual cost of delivering coal to consumers and argued that "coal consumers are deprived of the direct benefits of canal transportation." Boatmen along the Schuylkill Canal concurred. "The coal operator may claim the leasehold estate and the colliery," they argued in 1864, "but he is really an overseer, working for the Railroads at a per centage." Unless some action was taken to curb the influence of powerful carriers like the Philadelphia and Reading Railroad, the boatmen argued, the railroads would "control the market and may require the consumer to purchase at almost ruinous prices." The *Chicago Tribune* blamed the agents of anthracite carriers, who "made contracts with the Chicago coal dealers that they will sell no coal to be shipped to this city, unless it comes through them." As a result, the editors argued, coal was selling at a 200 percent markup in their city. "It is an outrage," the paper concluded, "which no respectable community ought to tolerate, and we trust something will be done to break the monopoly." In these three very distinct markets—New York, Philadelphia, and Chicago—very different actors arrived at the same conclusion regarding the source of high prices.[29]

In a more proactive move, some consumers banded together to form mutual coal companies that sought to mimic, not necessarily attack, the trend toward consolidation in the coal trade during the Civil War. Not surprisingly, Pennsylvanians led the way. In 1864, the incorporators of the Consumers' Mutual Coal Company denounced the soaring cost of coal, which it claimed was due to "the cost of labor at the Mines, the transportation changes, wharfage, or yard expenses, and profits of the operator and the retailer." "If the expenses can be diminished by the consumers working the mine through agents they have confidence in, and having it delivered directly to their doors," they argued, "the price can, of course, be so made less than the retailers price, and how far they can control the operation by which these advantages are to be secured." The idea behind the Consumers' Mutual Coal Company was simple: if coal was delivered to the yard in Philadelphia at seven dollars a ton and retailed at eleven dollars a ton, with carriage costs at only about fifty cents, why support the middlemen? Instead, this company—chartered under Pennsylvania's new liberal chartering regime—offered stock subscribers at ten dollars a share the ability to purchase ten tons of coal per year at net cost for the duration of the firm's charter, sixteen years. The price offered shareholders in 1864 was $7.50 a ton; any excess production would be sold at the market rate and the dividends distributed to shareholders for an even greater gain. "Every householder and business man in the community can participate in and should avail himself of its advantages," the prospectus crowed. Similar endeavors appeared in Boston and New York by the end of the war, although New York's Harmony Mutual Coal Company altered the structure of the cooperative venture to offer one ton per share at seven dollars per ton. The cooperative movement stretched down to the nation's capital, as Washington's Housekeepers' Coal Company offered stockholders the ability to "become directly connected with the mining and delivery of your own Coal." These cooperative companies, though often limited to a single anthracite mine, prospered as long as the retail price of coal remained high. By the end of 1864, the Consumers' Mutual Coal Company convinced 627 Philadelphians to subscribe.[30]

Once the war ended, the price of coal eased and consumers found relief. High coal prices, as it turned out, owed much more to temporary factors such as wartime demand and inflation than corporate manipulations of the market. Once the price spike subsided, though, consumers and users of coal still confronted a very different mineral fuel network dominated by large mining companies, well-integrated railroads, and dealers organized into coal exchanges. Each of these entities fought each other for market share—as the frustration of the Rockwell brothers in postwar Chicago suggests—but the sheer abundance of coal in the United States meant that no single actor or sector of the trade could control it. In an era of general industrial amalgamation, coal companies or unions, no matter how large and integrated across state boundaries, could not control prices. The interaction of political and geological factors in a few short years forged a national industrial network of mines, canals, railroads, retailers, and consumers that delivered cheap coal for home heating for the next century.

As a result, Americans burning coal in stoves now found their personal comfort to be directly dependent upon the nation's industrial economy. In 1874 journalist Charles Barnard tried to illustrate the complexity of this link to the readers of *American Homes* magazine in a three-part series entitled "From Hod to Mine," tracing the path of coal from his Boston hearth back to the mine. Barnard began by describing his trudge to the basement coal bin to fill his landlady's hod for the evening supply, only to find the bin nearly empty. After agreeing to secure more coal at his landlady's preferred dealer, Batchelder's, he traveled to Boston's Federal Street, "just where the endless line of coal-wharves extends in picturesque confusion in both directions." He struck up a conversation with a dealer who informed him that his annual business was sixty thousand tons—a substantial increase from when the dealer entered the business two decades earlier and found three thousand to be a "good year's work." Later that evening, while in the "glow of genial anthracite," Barnard "spread out [his] hands and wondered how many people, men, women, and children, had contributed to [his] comfort." He decided then to personally trace the route of his fire back to its source and to "see the whole business, from hod to mine, and make the personal acquaintance with every man or child who is in any way concerned in supplying a Boston fireplace or stove with good coals."[31]

Barnard began by securing passage on a coal steamer, or "tub," named the *Reading* to Philadelphia. After boarding, he observed the coal heavers unloading the vessel, a job so "severe" and performed at such a "furious speed" that "you and I would probably faint away in one half hour; and ten hours of such labor would lay us up for a month, and perhaps for life." To replace the weight of the coal, the *Reading* flooded its iron keelsons as ballast and was under way. Following a dark and stormy jaunt down the Atlantic coast, Barnard described the endless procession of vessels carrying coal and oil leaving Philadelphia, where the coal carriers stood out with their "deep sunk decks and grimy sails." Once in Port Richmond, the scope of the trade came into focus: "Twenty-one wharves in a row, thirty acres of coal yards, ten miles of railroad tracks, and a hundred and fifty thousand tons of coal." Before he even went ashore, Barnard watched the *Reading* begin loading for its return to Boston "on the early morrow." "Everywhere neatness, order, perfect system, and an air of grave propriety" permeated the docks, "as if the coal business was a very grim and solemn affair."[32]

Barnard secured passage on the Philadelphia and Reading Railroad for the final leg of his journey, a trip to Schuylkill County's anthracite mines. As the train entered the great coal terminal of Pottsville, Barnard noted that local residents seemed unimpressed by the "mountains that do stand around about give it a most romantic and picturesque appearance," speculating that "perhaps their souls are given to the contemplation of the beautiful in coal." From Pottsville the journey took him to Mount Carbon, then Tamaqua, and finally Mahanoy City, with its "deep valley, thickly wooded, and crowded with breakers and colliers, all actively at work." Barnard described the work on the breakers, with its conveyor belts and sifting screens. Here he described the human cost of the endeavor: the breaker boys hunched over the streaming coal and picking slate while an adult brandished a stick to discipline them. "What kind of citizens does Pennsylvania think these breaker-boys will make?" he asked. "Better burn slate in half the stoves in the country than drive school-children to such work with such a task master." Finally, Barnard descended into a three-hundred-foot mine shaft where he encountered "abrupt corners, crawling almost on hands and knees, through rough holes in the wall, climbing up steep, rough slopes, stumbling over masses of broken coal, sliding down banks of coal dust, splashing through puddles of inky water, and stumbling over interminable tracks and past countless cars" to get to the coal face. He concluded his journey by witnessing a powder shot close up and mingling among the men, "rough, grim with powder smoke, with hands hard with coal dust." "It is a long procession that ends with these miners," Barnard reflected. "Coal heavers, sailors, railroad men, breaker boys, miners,—we are debtors to every one of them."[33]

"From Hod to Mine" was intended to celebrate the complexity and labor intensity of the home heating network by the 1870s; Barnard's goal was to make his readers appreciate the hard work that went into keeping them warm. But with coal's rise to national prominence as a heating fuel, new problems began to arise. The use of coal in the urban hearth was efficient, but it also created a number of unanticipated problems with dirt and air quality within those households. The networks of coal production and distribution that emerged from the Civil War offered a host of new challenges as colliers, carriers, and dealers all struggled to control the trade. Coal was very useful in easing the scarcity of fuel in the nation's hearths, but even as it warmed American cities, some of the economic and environmental consequences of burning cheap coal began to become apparent by the late nineteenth century. Comfort came with a price.

4 How the Industrial Hearth Defied Control

ALTHOUGH THE coal-burning stove had conquered the home heating markets of American cities by the mid-nineteenth century, many Americans still struggled to master their stoves. Burning coal was cheap and effective in heating houses and rooms, but it was also dirty. Despite the best efforts of the breaker boys, dealers, and consumers, the anthracite or bituminous fuel that ended up in urban hearths contained impurities. With combustion, these impurities took to the air in the form of soot and ash or remained in the stove as "clinkers," as nineteenth-century Americans called the hard detritus that would not burn. Even anthracite, which was renowned for producing less smoke than bituminous coal, left a mess of ashes, clinkers, and soot after it warmed a room. As late as 1890, the English writer Alfred Pairpoint described a soot-covered American friend who had been "wrestling with his stove." "The stove needed a thorough cleaning with black lead," Pairpoint wrote, "and had, in the figurative language of our friend, 'to be wrestled with'; because, one would suppose, the stove didn't like it." This battle was constant. In addition to removing ashes and clinkers, users of coal-burning stoves needed to polish them regularly to prevent rust and cracks. This meant waiting for the stove to cool, mixing up a paste made of stove polish and water or vinegar, and laboriously scrubbing every surface of the apparatus. "Blackening" a stove in this fashion improved its appearance and lengthened its life, even as it wore down housekeepers. Stovepipes leaked smoke, dumped ash and soot into rooms, and often needed to be repositioned in order to work effectively. Fixing these wayward pipes, Pairpoint argued, became a time for American couples "when more harsh words were used on the occasion than any since the pair was united in wedlock." The process of heating the urban industrial home was indeed hard and dirty work.[1]

As Americans in cities burned more and more coal in fireplace grates and stoves, the work of maintaining the home heating system fell upon the women of the household. Battling the stove and its unruly pipes might involve the whole family at certain points, but day-to-day care of the hearth was considered a woman's work, as it always had been. The coal-burning stove made this responsibility even more dirty, demanding, and time-consuming. "This was a horrible messy job," Hazel Webb Daniel recounted of clearing her family's stove of ash and clinkers, and "it was always Mother who polished the stove." Working-class families placed the primary responsibility of maintaining home fires in the hands of wives and daughters in the decades following the Civil War. Even if a household could afford servants, the task of cleaning and polishing stoves almost always fell to female staff. In 1870 the *Ladies' Own Magazine* described a failed attempt to maintain a stove by an affluent protagonist, saved only by the appearance of an "Irish girl with a fan-shaped nose and wide mouth" looking for work. After being shown how to cool down the stove and mix the polish, the servant provided the writer with a critique of the high-end stove while her "tongue kept pace with the brush." "Yer stove's got too many figgurs on it," the servant offered. "It'll take a hard rubbin' to keep it clean." Such stories of toil were repeated again and again across northern cities as burning coal developed from an attractive and cost-saving option for home heating into a necessity of the industrial household. Although coal stoves saved families money on fuel, within the household they made even more work for women.[2]

Burning coal in stoves or grates reordered the ways in which urban Americans stayed warm and connected them to a complex economic system—creating in effect an "industrial hearth" to replace the traditional wood-burning one—yet also produced new challenges in the decades following the Civil War. This system might have seemed efficient in delivering cheap heat, but its sprawling nature and many participants made it uncontrollable and unpredictable. Increasing and intensifying the work of women within the household was one immediate impact for urban families. Health hazards, new types of fuel shortages, and conflict and disorder across markets for home heating fuel also tempered the benefits of mineral fuel. Coal might have been the first element of American industrial life, but by the 1870s and 1880s, several problems with the coal-burning stove system were becoming evident. Cheap stoves and abundant coal helped conquer the winter's bite, but the quality of air suffered both inside and outside of the home as a result. As mineral fuel became more ubiquitous, dealers and colliers alternately colluded and collided in order to gain the upper hand in its distribution network. The attempt to control the supply of coal, coupled with the constant threat of strikes in the mining region, tempered the enthusiasm for the nation's newly integrated system for distributing mineral fuel. These new environmental and economic challenges undermined the idea that burning coal offered an unqualified vision of progress for industrial cities of the North. The struggle to deal with these new challenges both within and outside of the household is the focus of this chapter.

Health and Home Heating

Coal-burning stoves did a good job of keeping rooms warm, but the quality of interior air suffered. This belief that stoves churned out cheap but unhealthy heat had deep roots. In the colonial period the Anglo-American tradition of the roaring wood-burning fireplace valued the "elegant" warmth produced by the open flame as it denounced the dry and stifling heat produced by German- or Russian-style closed stoves. Early American designs like the Franklin or Rittenhouse stoves attempted to blend the comfort of the open fire with the efficiency of an iron stove but failed among consumers because of their high price tags and among users because, quite simply, they didn't work. As the use of closed iron stoves increased during the 1820s and 1830s, this Anglo-American prejudice evolved into a critique of home heating's industrial character. In 1841, the *Boston Medical and Surgical Journal* published a lengthy article by a London physician titled the "Bad Effects of Breathing Impure Air" that criticized the public's need "for polished metals in almost every department of our domestic appendages, united to the interests of the furnishing ironmongers," who produced an artificial demand for iron stoves. *Scientific American* praised the "economy and comfort" of closed stoves but criticized their tendency to "give out an eye-smarting, throat drying something that is disagreeable, and cannot be healthy." Morrill Wyman, author of the influential volume *A Practical Treatise on Ventilation*, found that stoves used to heat rooms during winter produced "an uncomfortable degree of dryness, or even inflammation" and that "many persons experience a painful sensation in the chest" as a result of the dry air. Popular remedies such as placing pans of water on top of stoves to humidify the room could offer a temporary and imperfect respite, but many coal-burning families simply chalked up the stuffy atmosphere to the price paid for cheap heat. "The re-respired, roasted, ill-conditioned air of the dwellings even of the rich," Boston's Dr. Luther Bell argued in 1848, "is the result of a parsimonious economy, which strangely and absurdly exists as to this, in a thousand instances, where comfort and luxury make no other sacrifices to saving." Fuel economy, he argued, was a false economy that allowed for the "spreading missile of destruction, poisoned air."[3]

The idea that stoves actually poisoned the air gained momentum at exactly the same time that the coal distribution network blossomed. In the eyes of many reformers and physicians, the impact on the nation's health was dire indeed. In 1851 the fiery abolitionist Orson Murray published a lengthy jeremiad against stoves in his freethinking journal the *Regenerator* in which the "vitiated air of close stoves" were depicted as a national poison that affected "nine-tenths of all the houses in the northern states, whether belonging to rich or poor." "A person with his senses unblunted," the *Regenerator* reported, "has only to go into an ordinary unventilated room, heated by a stove, to perceive how at once, by the effect on the lungs, how dead, stifled, and destitute of all elasticity the air is." A New Hampshire physician wrote a long essay on the danger of stove-heated air in the *American Journal of Science and Arts*. While the open fireplace was inefficient but good for the ventilation of air, "modern ideas of comfort" required a more economical means of heat. The fuel efficiency of a stove, however, often produced "excessive" heat. If a room, for example, reached "summer temperature" of seventy-six degrees, it could provoke an "unusual dryness of the air-passages of the mouth and throat" for those inside and, for those leaving the room, an increased "risk of taking cold by the sudden check of the perspiration in the parts of the surface most exposed to the chill of the atmosphere." D. M. Dewey's 1852 pamphlet, *Heat and Ventilation*, made careful note of the noxious side effects of economical stoves for the poor. One physician, he reported, found "no mode in which the health and life of a person can be placed in more insidious jeopardy than by sitting in a room with its chimney closed up with such a choke-damp–vomiting stove." Saving on fuel thus had dangerous consequences. What was to be done? "If a red-hot stove destroyed, instantly, one man in every town daily, for a week," one sympathetic review of Dewey's pamphlet predicted, "there might be some salvation in the nation" from the plague of stove-heated air.[4]

The hot-air furnace—basically a large stove that sat in a basement and radiated heated air to upper floors via an iron piping system—made its first significant appearance in home heating markets of the 1840s. Furnaces were common in institutional settings such as churches, schools, and government buildings, but individual consumers saw a rise in these devices at precisely the same time that stoves became a ubiquitous—and messy—presence in households. Dealers in hot-air furnaces were more than happy to remind potential customers of the health and economic advantages of their product. Boston's Nicholas Mason, for example, denounced the use of stoves in cities as "the most pernicious [hazard] to health that can possibly be imagined" and argued that these devices were responsible for "the great proportion of consumption and bronchial disease in New England." Mason's hot-air furnace, when properly installed in a basement surrounded by brick and piping to transport the heated air to the rest of the dwelling was, of course, a stellar solution to the problem of closed stoves. Another Boston dealer in hot-air furnaces, Walter Bryant, admitted that faulty installation of his products could lead to the poor circulation of air but insisted that hot-air furnaces generally saved immensely on fuel. "If the head of the family will devote five minutes every morning after the fire is built to regulate the Furnace," he argued in 1861, "a steady heat all day will be secured, and a large amount of fuel saved." By enlarging the combustion chamber and removing it from close proximity to a dwelling's inhabitants, hot-air furnaces quite literally put the health risks of stoves out of sight for urban households.[5]

Although hot-air furnaces represented a significant advance in home heating technology by the time of the American Civil War, it was widely adopted only among affluent consumers. Like many changes in the method of warming houses, hot-air furnaces depended upon wealthy consumers to lead the way in the initial installation of the system. "Most of the enormous hot-air furnaces in use are placed in brick chambers," *Appletons' Mechanics' Magazine and Engineers' Journal* noted in 1853, "the construction of which constitutes an additional expense, such as can be bourne only by the wealthy." Walter Bryant's confession of the "difficulty of making servants pay proper attention to their management, in this and other particulars" suggests that his Boston clientele were drawn from the well to do. Much like the early adopters of coal in the 1820s or 1830s, the purchasers of hot-air furnaces in the 1850s and 1860s enjoyed the luxury of experimenting with their home heating systems. Since potential adopters of this technology needed both a large amount of space for the furnace and room to store three or four tons of coal, sales were quite limited. The popular opinion that coal-heated air was inherently unhealthy also slowed its adoption. Among antebellum tenants, the desire to retain control of their heating systems limited the appeal of hot-air furnaces to landlords. "The tenants whom we propose to accommodate are such as would, at all events, wish to keep their fires in their own rooms for cooking and other purposes," an 1846 report on rental housing in Boston found. Tenants "generally use[d] stoves," even though the "rooms [were] apt to be too warm" and lacked proper ventilation. These factors combined to keep hot-air furnaces, and central heating systems in general, on the margins of home heating markets.[6]

As households fought to remove smoke and soot from their interiors, more and more of the offensive materials accumulated in the shared urban environment. The anthracite-burning cities of the East Coast faced less air pollution in the nineteenth century, as "stone coal" burns relatively cleanly. But in cities closer to bituminous fields, like Pittsburgh, Cincinnati, Cleveland, Saint Louis, or Chicago, the quality of air suffered with the rise of mineral fuel. After the Civil War, travelers could not help noticing the atmospheric impact of cheap local coal. "Smoke pervades every house in Cincinnati," the journalist James Parton wrote 1867; it "begrimes the carpets, blackens the curtains, soils the paint, and worries the ladies." One English visitor claimed that Cincinnati's "densely occupied and smoke-enveloped streets" gave its central business district an aura of "premature antiquity." Cleveland's air fared little better in the eyes of visitors. "If one returns from an hour's walk he comes in appearing like a respectable mulatto," one visitor from Sandusky, Ohio, wrote, "and his nostrils are saturated with soot." A correspondent to the *Boston Daily Advertiser* wrote in 1867 that "St. Louis abhors whiteness—as nature abhors a vacuum; those perennial showers of soot and that all-pervading smoke takes the edge off of cleanliness 'right smart.'" Pittsburgh, though, seemed the dirtiest of all. Residents' proclivity to build coal fires provoked complaints as early as 1804, and James Parton noted during his visit there in 1868 that "dainty and showy apparel" could not thrive in the atmosphere. One travel guide written by William Glazier in the 1880s ventured that, at its best, Pittsburgh was a "smoky, dismal city" but at its worst, "nothing darker, dingier, or more dispiriting can be imagined." In 1892—nearly a quarter century after Parton's *Atlantic Monthly* article—little had changed to improve Pittsburgh's air quality. "In all our travels over a large part of the world," Alexander Craib wrote, "it is the sootiest, smokiest, dirtiest city we have ever seen." Indeed, he concluded, Pittsburgh was "a place enveloped, defiled, and famous in smoke."[7]

When confronted with the problem, some residents of smoky cities defended them for various reasons. The common wisdom about the necessity of smoke stubbornly prevailed for much of the nineteenth century. The notion that coal smoke was antiseptic and therefore had a cleansing effect on the lungs was one common nineteenth-century claim. Even if smoke and soot induced coughing and wheezing, in the long run it was healthy for residents as a disinfectant, the argument went. One Pittsburgher told James Parton that "the smoke of bituminous coal kills malaria and saves the eyesight." Another defense held that economic progress required smoke. The *Cleveland Herald* reported that many of its citizens found their "sooty garb as giving evidence to their prosperity." Dirt and smoke from urban hearths and factories were unseemly but necessary symptoms of industrial affluence. "The people must have fuel," the *St. Louis Globe-Democrat* argued in 1881, "and the cheapest fuel is coal; and it is not easy to burn coal without producing the annoyance of sooty smoke." Although Progressive Era reformers eventually dispelled these myths about the healthfulness and economic necessity of coal smoke, abatement programs found them serious impediments in the late nineteenth century. It may seem inconceivable to modern-day Americans, but it took a great deal of energy, time, and money to persuade the residents of these cities that smoke could be reduced. And when, after much lobbying, smoke abatement programs did appear in northern cities, most ordinances targeted manufacturing and commercial chimneys and exempted domestic hearths from regulation. As a result, the quality of city air in the era of the industrial hearth remained poor for some time.[8]

The incredible growth of America's largest cities exacerbated these environmental problems, even in areas that burned relatively clean anthracite. By 1870, for example, New York and Brooklyn combined had 1.3 million residents—a nearly four-fold increase from their population in 1840. Philadelphia grew to 674,022, an incredible increase of 719 percent over its 1840 population. But most impressively, Chicago emerged as one of the top industrial centers in the United States by 1870, boasting a population of nearly

300,000. Considering that Chicago had only 4,470 residents in 1840 and little more than 112,000 in 1860, that city's spike in population made it one of the fastest-growing urban areas on the planet. The increase in the North's urban population reflected the interrelated influence of industrialization and immigration; the vast majority of new arrivals in the cities of postwar America were there to work in the burgeoning factories, shops, and foundries of the manufacturing sector. Such rapid growth over a few decades placed a massive strain on each city's existing stock of housing—the three- or four-story buildings that made up the core of antebellum cities slowly and unevenly became tenements, or multifamily dwellings characterized by crowded conditions, high rents, and squalid living conditions. In some cities, existing buildings were converted into tenant housing; in others, new "railroad tenements," narrow buildings that had up to sixteen rooms per floor, appeared. Whatever their origin, the tenement houses of mid-century American cities packed in more and more residents. By 1850, Boston's and New York City's tenant buildings averaged sixty-five residents. In 1864, a survey of the latter city's tenements found 486,000 New Yorkers living in 15,511 tenement houses with population densities along the Lower East Side of Manhattan reaching more than 300 people per acre.[9]

The story of New York City's struggle with the "tenement problem" illustrates the struggle to come to grips with this new style of urban life and in particular the difficulty of tenement residents to secure cheap and efficient home heating. By the end of the Civil War, middle- and upper-class New Yorkers viewed the conversion of industrial warehouses and older single-family residences into tenements as creating more than simply unpleasant housing conditions for working-class families. The urban core of northern cities underwent a massive growth in population, fueled by European immigration and the continued migration of rural Americans to cities. As they packed into the dense, cheap housing, these new urban residents struggled to attain even the most basic comforts of life. The rise of crime and disease in these crowded urban areas—by this time likely referred to by the relatively new British term "slum"—threatened to overwhelm city services in public health and order. In 1867, New York passed a "model tenement law" that attempted to regulate the construction of multifamily dwellings and particularly to reverse the tendency of these structures to have unlit, garbage-strewn passageways with poor ventilation and light. The 1867 legislation mandated proper air flows, access to windows, and sewage disposal in each new dwelling but left home heating to the discretion of the tenants, dictating that each tenement "shall have adequate chimneys running through every floor, with an open fire-place or grate, or place for a stove, properly connected with one of said chimneys, for every family and set of apartments." The 1867 law offered only sporadic relief from the problems facing urban slums, so the editors of the *Plumbing and Sanitary Engineer* magazine offered a prize for the best improvement in the design of tenements in 1878. The winning entry was the "dumbbell tenement," which improved ventilation and light for multifamily dwellings but assumed that residents would use a fireplace, grate, or stove for home heating. When New York passed yet another Tenement House Act in 1879, it adopted the "dumbbell" plan and, along with it, the notion that model housing for the industrial working class needed to provide an outlet for soot and smoke but virtually nothing else. New York's tenants were on their own when it came to home heating.[10]

Even though the residents of tenement housing retained control over how they heated their homes, this was likely a hollow choice faced by tenants in postwar cities. The need to make small-scale purchases of fuel drove the price up for those who could least afford it. And even if they could purchase coal by the ton, where would families living in tenements store it? If New York, the center of housing reform, could not improve the situation, then problems such as fuel storage would continue to plague rental housing in cities. An 1870 study of working-class households found that Boston's model tenement house on Kneeland Street had twenty compartments for wood or coal fuel in the basement but that most other renters lacked adequate space to store heating fuel. Inspectors from the Massachusetts Bureau of Labor Statistics reported that coal was kept in closets, in cupboards, or under stairs; all of these locations suggest that only a small amount of coal could be stored. One worker told inspectors that "one in ten can put in a winter's coal" ahead of the season, and when Boston's harbor froze up, dealers demanded payment in advance. Poor consumers there bought coal by the "peck," an informal measure that was about twenty pounds. A Massachusetts survey of 1870 estimated that coal secured in this fashion cost about eighteen dollars a ton—about four times as much as the price charged for larger purchases. More often than not, renters fended for themselves, and home heating took a backseat to tenement reforms designed to improve health and sanitary conditions in urban slums. As late as 1889, Boston's city code required that every tenement building "have adequate chimneys running through every floor, with an open fireplace or grate, or place for a stove," along with the facilities to collect noncombustible waste.[11]

In addition to housing problems, concerns over the health and well-being of families in a coal-burning home became a common theme in postwar treatises on domestic conditions. The Boston physician George Derby led the way with his brief but influential 1868 pamphlet entitled *An Inquiry into the Influence of Anthracite Fires upon Health*. Derby first went to great lengths to defuse some commonly held notions about air heated by stoves, grates, or furnaces that burned anthracite. The notion, for example, that heated iron drew moisture out of the air at a rapid rate did not hold up under his experimentation, nor did the idea that the by-product of combustion, "carbonic oxide" (carbon monoxide), was a harmless gas. Dispelling these two myths was essential for understanding why anthracite heat could have the effect of a "narcotic poison" on many individuals who burned coal. But Derby's most influential contribution to this growing discussion was to blame the imperfect construction of stoves, chimneys, and furnaces, which allowed carbon monoxide to mingle with breathed air. Dryness, therefore, was not necessarily a problem, but the chemical composition of the air was. In fact, he argued that "carbonic oxide gas is the chief and probably the only cause of the unpleasant sensations and injurious influences so commonly associated with the combustion of anthracite coal." Derby offered an expensive solution that undoubtedly raised the cost of burning coal: make sure that all pipes and fittings are airtight, keep a fire lit in the stove or furnace at all times to ensure proper ventilation, and keep the fire relatively low so as to discourage carbon monoxide's spread. Using a wood-burning furnace or steam to heat homes, he concluded, would be the best solution of all, if not the most economical. Derby did not continue to study heat and ventilation, but in 1869 he became the first secretary of the Massachusetts Board of Health. As a champion of using the scientific method and statistical evidence to actively combat epidemics and disease, Derby launched a study of Boston's tenement homes as a response to findings of the Massachusetts Bureau of Labor Statistics that houses of "decency and comfort" were rare for the city's workers.[12]

Ventilation and heating challenges to more affluent users of coal attracted the attention of Catharine Beecher and Harriet Beecher Stowe, the nation's leading writers on domesticity. By 1869 the need to reconcile the new preference for coal-burning devices with the reduced quality of air drew Beecher and Stowe's ire in *The American Woman's Home*. Following a lengthy explanation of how human respiration works and extensive medical testimony as to the necessity of clean air, Beecher and Stowe argued that "civilization has increased economies and conveniences far ahead of the knowledge needed by the common people for their healthful use" and that the "heating and management of the air that we breathe is one of the most complicated problems of domestic economy." The majority of Americans, they argued, "owing to sheer ignorance, are, for want of pure air, being poisoned and starved." *The American Woman's Home* drew upon, of all things, coal mine ventilation techniques as an example of how a lit fire draws air into a closed space; an adapted metal tube in the domestic setting substituted for the furnaces that colliers placed at the bottom of the mine shaft to pull fresh air in from the surface. They compared favorably the ways in which both domestic and industrial systems drew clean air in from the outside while flushing the "carbonic acid" and foul air through the top of the pipe. Even the poor, with their "stinted store of fuel," might benefit from the application of coal-mining practices to the coal-burning home. Beecher and Stowe provided their readers with the latest stove designs for both cooking and heating homes in *The American Woman's Home* but reserved judgment on the wisdom of substituting fireplaces or stoves with furnaces, as the latter tended to dry out the air at disturbing levels, heated rooms unevenly, and, most disturbing, produced more warmth than was necessary. Believing that cool air fueled the brain more efficiently, they concluded that "the injected air of a furnace is always warmer than is good for the lungs, and is much warmer than is ever needed in rooms warmed by radiation from fires or heated surfaces." In their plan for economical "cottages" built with an eye toward comfort and economy, Beecher and Stowe recommended a construction plan that favored central chimneys to conserve heat. But rather than a central furnace or a network of stoves, they championed the old-fashioned Franklin stove, with its "radiating warmth and cheerful blaze of an open fire." In the future, they hoped that a "benevolent and scientific organization could be formed" to inform the public of the most economical and healthy stove designs, thus saving the nation "both millions of money and much domestic discomfort."[13]

Much of the prescriptive literature on domesticity idealized a world of "separate spheres," in which women and men labored separately in private and public settings. The coal-burning household, however, reflected a different reality, in which men and women were both connected to the wider networks of home heating. *The American Woman's Home* offered the latest theories in scientific domestic economy, but its prescriptive elements that emphasized a return to rustic ideas of comfort seemed out of touch with the realities of industrial society. Simply put, Beecher and Stowe failed to realize the many architectural and economic constraints of less affluent households for home heating reform. Franklin stoves could hardly help those families, so this was terrible advice for the majority of Americans. Other domestic guides aimed at home economy carried less celebrity punch than *The American Woman's Home* but echoed the same message in terms of home heating. Joseph and Laura Lyman's *Philosophy of Housekeeping* (1869) argued that the "heat of anthracite is more concentrated, and often greater than the comfort of the apartment," even as it remained the most economical fuel. Except for houses in the "extreme north," the Lymans also found the Franklin stove to be the perfect blend of aesthetic aspects of the open fire and the economy of coal. In those regions that suffered from the most frigid temperatures, the hot-air furnace was an attractive option, even though it "requires a very liberal consumption of fuel." Innovations in home heating designed for comfort found themselves at odds, often quite explicitly, with fuel economy.[14]

The failure of domestic handbooks to recognize the limitations of many urban families in home heating reflects a larger disconnect from the reality of the industrial household. Instead of sharpening the distinction between publicly oriented male purchasers of mineral fuel and homeward-looking female builders of coal fires, industrial households entangled with national fuel markets blurred these lines even further. The desire among many writers on domesticity to return to the good old days of open fireplaces and roaring wood fires offered the same unrealistic goal for most urban families in the 1860s and 1870s as Peale's fireplace or Pettibone's furnace did in the 1790s.

Keeping the Industrial Hearth Lit

As the nation's stoves, furnaces, and grates required a large, steady supply of coal, both men and women in households of all varieties were involved in the process of securing adequate heating fuel, maintaining steady levels of heat and ventilation, and keeping their households clean. Within the home, some of these tasks were divided by gender, but keeping warm in the industrial city involved the constant attention of both men and women. Watching the pace of fuel consumption, deciding whether to lay in a season's supply and when to time that decision, and assessing which coal burned best were all tasks that demanded the attention of both users and consumers, women and men, within the industrial household. These households depended upon a retail distribution network for mineral fuel that could promise a consistent and affordable supply. As urban users and consumers of coal became further enmeshed within the postwar industrial economy, they became dependent upon the constant flow of coal into American cities.

A critical actor in this industrial play was the coal dealer, who served the all-important task of keeping urban households constantly connected to the nation's coal network. As distributive agents in the fuel commodity chain, urban coal dealers served a vital purpose but found themselves at odds with both their customers at the consumption end and their suppliers at the production end. By the 1860s and 1870s, coal dealers were some of the most influential—and notorious—retailers in American cities as they negotiated a complex new economic landscape filled with aggressive coal company agents, erratic supply and demand for their product, brutal price competition, and frustrated and angry consumers. While women struggled to keep the fires burning within the industrial household, men encountered similar difficulties in mastering the retail network designed to ensure a steady flow of mineral fuel from the coal regions to the cities.

Although they were a familiar presence in the urban landscape by the time of the Civil War, coal dealers varied in both size and scale. Many of them were quite modest operations with only a few employees and limited space (fig. 4.1). Large dealers commanded massive operations that purchased coal directly from coal companies, unloaded their product at their own wharf, and employed a small army of delivery carts to distribute their coal to both long-term and one-time customers. These carts dumped their cargo either on the street in front of their customers' homes or into a "coal hole" for storage. By 1855, Boston's Board of Aldermen declared these coal holes to be urban nuisances and demanded that they be covered by a series of patented covers. Seven years later Bostonians were told by city authorities to keep their coal holes at less than eleven feet in depth and to cover them with a "substantial iron plate, with a rough surface, to prevent accidents." Two decades later, a New Yorker denounced the still "ordinary custom of dumping the coal upon the sidewalk" as a "most unhandy and unclean arrangement." Smaller dealers or grocers who dealt in coal worked out of smaller lots, often distributed throughout the city, from which they sold coal in tons as well as smaller amounts—perhaps by the peck, bushel, or even bag. They served the growing population of urban households who could not afford to purchase coal for the season or store a large supply. Whether onc kcpt his or her fuel in a cupboard, under the stairs, or in a spacious coal hole, consumers likely secured it from a retail coal dealer.[15]

The variety of coal flowing into American cities from various sources allowed postwar coal dealers to keep prices relatively low, even as they complained bitterly about securing coal at wholesale prices low enough to turn a profit in retail markets. Urban coal yards secured their season's supply by arranging shipments between suppliers—often coal-mining companies, railroads, or wholesale merchants—and dealers. Unlike large manufacturing companies or railroads, both of which sought to lock in fuel prices over a long period of time, coal dealers generally did not sign long-term contracts with colliers or carriers of coal. Bargaining on prices was a freewheeling affair. Being a coal dealer was not for the fainthearted, as each season brought intense negotiations. In 1850, the merchant and shipper of Pennsylvania anthracite Supply Clap Thwing made it clear that Boston's coal dealers were tough negotiators on price and that his supplier, the Forest Improvement Company, had not introduced its best coal to that market. The dealers, Thwing complained, will not "give a cent more for it than for coal they have tried and I have never been able to offer them sufficient inducement to increase our sales as much as we wished in this branch." In Chicago, sales agent John Kirk represented several anthracite colliers in Pennsylvania but could not wedge his product into the yards of local dealers, who wanted him to lower prices. "We should in self defense be compelled to open out a coal yard here for ourselves," Kirk lamented in 1867. Providence coal retailer George Bowen kept a close eye on prices, which he could negotiate directly from coal-mining companies or could secure, via agents, at periodical coal auctions at urban wharfs in New York or Philadelphia. In 1869 Bowen enclosed a check for one payment to a wholesaler with a note complaining that "it is rather hard to pay so much more than expected" and a few weeks later demanded a lower price from another by asserting that "all expect coal to fall after the Sale in New York." As more coalfields came into play, and as more carriers competed for traffic between mining regions and urban centers, coal dealers developed a reputation as hard-nosed businessmen operating in an extremely competitive environment. "The coal dealer, by the system which now appears to be his second nature," one former collier noted in 1887, "is probably the most important factor in keeping the whole coal trade steady."[16]

Figure 4.1. A. McHale's Lehigh and Schuylkill Coal Yard, 1885. The coal yard was the basic retail outlet for the nineteenth-century trade. This image demonstrates the small size of many coal yards, which in this case was operated by a few proprietors with a single cart for delivering coal to urban customers. Several of these small coal yards were scattered over the typical nineteenth-century urban landscape, where they competed fiercely on price. Library Company of Philadelphia

It might make sense for coal dealers to collaborate with each other to set prices, but the easy-entry and easy-exit nature of their business—a common trait among urban retail trades in postwar cities—frustrated such attempts. The coal exchanges that arose during the Civil War remained in place and attempted to offer some informal regulations on the trade by standardizing delivery methods, ensuring uniform weights, and announcing the price at which coal *should* be sold across their respective cities. As prices subsided from their Civil War peaks, coal dealers found themselves operating in a fiercely competitive market and earning thinner profit margins. In Chicago, for example, the Citizens' Protective Fuel Company promised to sell coal at only fifty cents per ton above cost. While these ventures were popular responses to wartime spikes in prices, private coal dealers were able to undersell cooperative ventures and drive them out of business. The managers of the Citizens' Protective Fuel Company didn't help by raising prices to their subscribers at the height of the winter season of 1868, after they had promised to lock in low rates. "No reasonable man would ask or expect to sell coal at $10 per ton," the directors argued, "which is costing over $13." Such claims drew accusations that the Citizens' Protective Fuel Company showed the "indices of a swindle," in the words of the *Chicago Tribune* and that subscribers seduced by the promise of low prices were "dupes," tricked by an "injudicious or knavish" management. Cooperative attempts to undercut private dealers failed by the late 1860s, and Chicago fuel markets became the province of private industry. Yet, even with the Chicago Coal Exchange attempting to coordinate supply, costs, and prices, competition between dealers remained fierce. "While there are countless small dealers who can undersell," the *Tribune* reported in 1871, "there are comparatively few larger dealers who are anxious to keep price up to the standard fixed by the dealers at their periodical meetings." So when small dealers undercut the Coal Exchange's suggested retail price and sold at $8.50 a ton, two of Chicago's largest dealers announced that they would sell at $7.50 a ton, triggering "joy and wonderment" among consumers for a few days. Eventually prices swung back into line with the expectations of the large dealers, but not before many small dealers suffered huge losses. In postwar Chicago, then, coal dealers priced cooperative alternatives to private retailers out of fuel markets but at the same time created a fiercely competitive price environment for their business.[17]

In addition to monitoring prices citywide, urban coal dealers also needed to balance the demands of their customers with the available supplies of coal. As anthracite emerged as the leading home heating fuel in northeastern cities, for example, dealers there advertised variants based on their location, color, or the name of the company that mined it. So at various points, dealers negotiated directly with colliers or wholesalers for shipments of "Shamokin White Ash," "Lykens Valley Red Ash," "Locust Mountain," "Franklin," or "Lorberry" coal that would be broken down in sizes described as "egg," "lump," or "nut." In 1866, the *Chicago Tribune* attempted to provide an inventory of the city's coal dealers. Aside from some small dealers and a few "adventurous spirits" who sold a ton here and there, reporters found at least twenty-two retail dealers, with stocks ranging from a few hundred tons to well over 50,000. The total amount of coal on hand in Chicago reached nearly 340,000 tons and varied from bituminous coal from Ohio's Mahoning valley (35%), Pennsylvania anthracite (30%), Illinois bituminous (21%), Pennsylvania bituminous (7%), and varieties of "Second Grade Bituminous" (7%).

Of course, consumers often put their trust in dealers to provide the actual product. In 1872, *The Coal and Iron Record* warned against the "scalawag" dealers of New York, Boston, and Philadelphia who offered coal beneath market price and "pretend[ed] to sell you the best Sugar Loaf Lehigh coal and make no bones in sending you some *mongrel* coal that is unknown, in its place."[18]

Dealers then balanced quality control with price in order to place their orders with coal companies or wholesale merchants. In the winter season of 1869, Providence's George Bowen badgered his wholesale contacts for high-quality coal. "The Cargo of Locust Mountain is the poorest coal I have had for years" and wouldn't sell, he wrote to one firm. "Your Henry Clay is higher than Lorberry and we get 50 cts. per ton more for it," he wrote to another. Bowen also complained of a shipment "much afflicted with stone," which was a common complaint that dealers relayed to shippers. Even as mining companies employed an army of small boys to pick stone and slate from coal, some waste material inevitably made it into shipments bound for urban markets. Dealers like Bowen used this common wastage as a negotiating tactic. "I have taken a great deal of pains to install to you the damage to me of a poor White Ash Coal," he wrote in October 1869. "I won't have another like it to sell my customers at any rate, it will do me great damage." "The last cargoes has been very stony and caused much complaints," Bowen wrote when ordering two hundred tons of Lorberry coal from the Philadelphia firm Blakison & Graeff. "I think you will have to be more careful not to send slate if you wish to have your Coal sell quick." Since a dealer's reputation, and thus his future, often depended upon customer satisfaction, quality control became an important aspect of the retail home heating fuel business in postbellum cities. "Quality should be the prime consideration," *Saward's Coal Trade Journal* argued in 1874, "and if the price of a superior quality of any description of merchandise is higher, you may be sure that it is the cheaper article in the end."[19]

Even as the bare-knuckle competition in fuel markets benefited fuel consumers, criticism of coal dealers in postbellum cities advanced on other fronts. The ways in which dealers measured coal, for example, drew intense scrutiny from the public (fig. 4.2). Since 1853, coal dealers had engaged in the common practice of buying "long" or "gross" tons of 2,240 pounds from wholesalers and coal companies but selling "short" or "net" tons of 2,000 pounds to their customers. This practice began in Philadelphia, where common complaints about slate and rock in coal shipments allowed dealers to account for wastage, which they argued made up more than 10 percent of most shipments. Over the years, however, urban consumers complained bitterly about the ways in which dealers shortchanged them on weight. In New York, for example, unscrupulous dealers employed delivery wagons that held between eighteen and nineteen hundred pounds; when a short ton was measured and poured into these wagons, the coal that flowed over the side remained at the yard. An 1869 investigation of sixteen dealers discovered that only two purchased loads actually reached two thousand pounds in weight; two of the samples were more than two hundred pounds short. Six years later in Philadelphia, a sample of six short tons from separate dealers ranged from 1,691 to 1,893 pounds in actual weight. The *Chicago Tribune* argued in 1874 that small dealers had more of an incentive to cheat, as a shortage of one hundred pounds per ton meant a fifty-cent profit on each short load that left the yard, and thus "the sooner it is understood that your neighborhood petty coal merchant swindles you inevitably and of necessity, the better it will be for coal consumers." When dealers broke down sales to smaller amounts to accommodate less affluent customers, the price per pound increased considerably, as did the incentive to cheat consumers with lightweight shipments. By the late 1870s, trade journals called for standardizing the retail sale of coal in regulation two-hundred- or one-hundred-pound sacks, so as to "do away with the pernicious practice of selling coal by the peck or bushel, by which the poor are daily swindled," yet nineteenth-century dealers continued to enjoy a great deal of latitude in the weighing of their product.[20]

By the 1880s, then, many coal dealers found themselves in an unenviable position. On the one hand, cutthroat competition reduced profit margins; the strong incentive to cheat consumers often led to public outrage against their trade. Coal dealers eventually saw federal, state, and local officials step in to regulate their trade during the twentieth century, but during the nineteenth century—the period of greatest change in home heating fuels—the trade remained without much public oversight. Municipalities licensed dealers but hardly demanded large sums of money or a tangle of regulations in order to retail coal. Consumers could use public scales but did so on their own time and at their own expense. Most cities continued to employ coal inspectors at wharfs and railroad yards, in the same tradition as the wood corders of the Early Republic. This regulatory system, however, did not intervene in the market relationship between retail dealers and consumers, which had become the main problem in postwar heating fuel markets. Delegates to New

Figure 4.2. A coal shipment at the scale. One of the bitterest complaints among customers was that dealers delivered light loads. Cities maintained public scales in order to restore faith that a delivered ton actually weighed two thousand pounds, but few customers took the time and effort to use them. The dealer's scale, which may or may not have been precise, usually sufficed. *American Agriculturist* 24 (1865): 200

York's state constitutional convention in 1868 debated the need for regulation in the retail coal trade. "What we want," one delegate argued, "is that there shall be a proper officer who shall be umpires between the buyer and the seller. . . . I believe short weight and short measure to be a very general abuse in New York." The counterargument placed the blame on greedy consumers, who were probably "seeking a seller at a short price." The competitive market in coal, in fact, placed informal limits on unsavory coal dealers, as honest ones "upon the merest pecuniary consideration, cannot afford to cheat in the matter of weight or sample." In the end, New York's coal dealers escaped state oversight. By 1876, *Hall's Journal of Health* complained bitterly about the pliant standards employed by the city's coal dealers. "It is perhaps known to few, that no coal dealer in Gotham ever, by any possibility, sells a lawful ton of coal," the editors argued, "although he is very clear of purchasing less than a lawful ton of twenty-two hundred and forty pounds, or twenty-eight bushels, each bushel being eighty pounds."[21]

Some states did pass laws to protect consumers, but they had a limited impact on the day-to-day retailing of coal. In 1871 Pennsylvania's legislature, for example, set the official weight of a ton of anthracite coal at 2,240 pounds and provided for three inspectors in Philadelphia to inspect carts and wagons used by the city's coal dealers. If an inspector suspected a cart of being light and was within four hundred yards of the coal yard, he could demand to weigh it. Dealers caught selling short weights could be fined fifty dollars. In theory this law gave the state oversight of the trade, but Pennsylvania's 1874 constitution discontinued state-level inspections and measuring. Philadelphia continued to appoint coal inspectors, but three officers regulating a trade in which railroads regularly shipped between 200,000 to 500,000 tons to Philadelphia every month by 1873, and in which annual production was estimated to be about 25 million tons annually, was nearly impossible.[22]

Instead of seeking public regulation, nineteenth-century dealers worked through coal exchanges or other organizations to sort out the problems of their trade. The Philadelphia Retail Coal Dealers Association (PRCDA) offers one example of how the dealers themselves confronted these issues. To counter the accusation that dealers were selling lightweight tons, the PRCDA required each member to submit reports to its Committee on Weights and Transportation. In October of 1880, the PRCDA issued a confidential circular to its 168 members asking them to advertise prices of the best-quality coal clearly, to offset "numerous advertisements of low price coal fictitiously set forth, as 'best Lehigh' whereby the public are deceived and led to believe that they are being imposed upon by the reasonable charge of honest dealers." A year later, the PRCDA praised the "greater uniformity of prices at retail during the past year," which was no doubt a result of claiming roughly 90 percent of Philadelphia's retailers as members. By 1883, the PRCDA increased its membership to 185 firms but noted that the influx of new dealers had hampered their efforts to secure an "improved margin of profit" and led to the persistence of "local evils" among nonmembers. But as difficult as incorporating all of Philadelphia's coal dealers might appear, the PRCDA saved its most vigorous criticism for coal-carrying railroads. "In these days of corporate power and monopoly," the PRCDA thundered in 1882, "there is no protection for the individual against the aggressions and abuses naturally growing out of such power, except by the united action of all interested, and that action must be persevering and vigorous." By redressing the two major consumer complaints against their business—quality control and standardized weights—organizations like the PRCDA hoped to restore consumer confidence in the trade. Railroads, they consistently argued, made that task impossible. Nonetheless, as the former collier Charles Miesse wrote in his 1887 handbook for the trade entitled *Points on Coal*, consumers demanded more and more of their coal dealers. "The consumer embraces a wide range, that may be described as the huge monster that employs legions to feed him," Miesse complained, "while he lashes his tail in their faces, and stares at them with astonishment and anger."[23]

Were coal dealers simply passing the blame? Eventually the drive to control prices expanded to include the carriers of coal. In most bituminous coal regions, such as the Illinois and Ohio fields that served large cities like Pittsburgh, Cleveland, Cincinnati, and Chicago, dealers could draw from lake, river, canal, or railroad traffic—in the case of Chicago, all of them. Bituminous coal prices thus tended to remain low and competition sharp, as railroads began to control the regional flows of bituminous coal from disparate fields to major cities. Bituminous colliers responded by attempting to slice both wages—thus running afoul of their miners—and prices, which narrowed their profit margins even more. By setting transport rates and consuming more than a quarter of all the bituminous coal mined in the United States for their own energy needs, railroads emerged as the major player in bituminous markets. Home heating fuel markets did not suffer much from this desperate battle among miners, colliers, and railroad executives; the smoke and ash produced by bituminous coal made it less preferable for domestic markets, even as American industries became more and more dependent on soft coal as an energy source.[24]

Cornering the Market on Heat?

Pennsylvania anthracite, however, remained in high demand regardless of cost. Its reputation as a superior home heating fuel—strong, clean, and familiar to most urban consumers—made it indispensable. Whereas bituminous fields appeared throughout the nation, the concentrated nature of anthracite fields tucked away in the mountains of eastern Pennsylvania offered another incentive for controlling the trade. Although the three major anthracite regions had been served by a competitive system of canals and railroads since the 1850s, large consolidated firms seemed poised on the brink of monopolizing the trade by the 1870s. Pennsylvania's Civil War legislature repealed long-standing policies that separated carrying and mining privileges; by the 1870s railroad companies responded by attempting to use this authority to monopolize the trade. No railroad company symbolized this battle more than the Philadelphia and Reading Railroad (PRR).

The PRR formed in 1833 to link its two eponymous cities and compete in the anthracite-carrying trade. When it began operation in 1842, the PRR found competition for the carrying trade to Philadelphia dominated by the water traffic of the Schuylkill Navigation Company. Through aggressive rate reductions that put the PRR on shaky ground financially—a trend that would characterize the firm throughout its history—the railroad captured the lion's share of the carrying trade by the Civil War. The wartime strikes and work stoppages that plagued Pennsylvania's anthracite fields interrupted the long-term efforts of the PRR to dominate the trade. But as state and federal officials undermined the nascent union movement in anthracite country, the PRR engaged in an aggressive campaign to grow its business by the late 1860s. In 1868 the PRR entertained ten coal dealers from Boston with an all-expenses-paid junket to visit the railroad's anthracite mines in Pennsylvania. While the dealers sipped complimentary Blue Seal Johannisberger, the PRR's vice president, John Tucker, told them that they "had already paid for it, and he intended they should do it again." On their way back to Boston, with a box of cigars courtesy of the PRR under their arms, the Boston dealers then toured the Delaware and Hudson's operations in Carbondale. Cultivating retail dealers offered one way to expand, but the PRR found controlling the supply of coal to be much more lucrative. The hurdles to attaining that goal, though, were quite high. Although some corporate consolidation had occurred during the Civil War, Pennsylvania's anthracite fields still saw a patchwork of small- and large-scale colliers seeking to raise as much coal as possible, which of course resulted in lower prices and even thinner profit margins. In 1868, a new union called the Workingmen's Benevolent Association (WBA) formed in the anthracite regions under the leadership of Irish immigrant John Siney. The WBA grew rapidly to include 80 percent of anthracite miners and laborers, and by 1869 the union forced local colliers to accept a sliding scale that linked wages to the price of coal. The WBA also staged strikes explicitly designed to lower the supply of coal and raise prices. With the sliding scale in place, labor would benefit from the stabilization of the trade much more than the carrying corporations would. By 1870, the PRR's dynamic new leader, Franklin B. Gowen, faced long odds in securing the company's future.[25]

Gowen's plan to tame the unruly anthracite trade had three basic goals. First, he wanted the PRR to acquire its own coal mines, either owning them outright or leasing them from small-scale operators. Second, Gowen sought to break the power of labor unions in the anthracite fields. Finally, the PRR would organize a cartel in the anthracite region that could set the price of coal and thus end the ruinous competition so endemic to the trade. Gowen knew firsthand that the coal business was rough. His Mount Laffe colliery failed in 1859 after a mine fire and flood destroyed it. But more than the physical challenges of mining coal, Gowen saw the atomization of the trade as its main problem. Between 1833 and 1875, for example, more than a thousand collieries were opened in Schuylkill County alone. More than half of these collieries failed after their first year of operation, and only 6 percent made it into their second decade. Beginning in 1868, first with his own money, then with investors', and finally by using the funds of the PRR, Gowen aggressively acquired coal lands, eventually organizing them into a PRR-owned subsidiary company, the Philadelphia and Reading Coal and Iron Company. By 1871, this firm owned more than one hundred square miles of coal-producing land; four years later it controlled one-third of the southern Pennsylvania anthracite field. All in all, the PRR acquired more than one hundred thousand acres of prime land at an estimated cost of $40 million between 1871 and 1875. By controlling nearly one-third of the anthracite shipments to Philadelphia, Gowen tried to reconfigure many of the long-standing practices of the trade. In 1873, for example, he eliminated a 5 percent discount traditionally offered by carrying companies to dealers to account for dirt and slate. That same year, Eugene Borda, the president of the Philadelphia Coal Exchange, accused Gowen of creating a "gigantic monster" that threatened the interests of both the anthracite coal region and Philadelphia's four hundred retail coal dealers.[26]

Gowen then turned his attention to managing the competition between anthracite carriers. For the first half century of the trade, a politically mandated separation between mining and carrying coal offset the ability of any one company to take advantage of the compact nature of Pennsylvania's anthracite fields. By the late 1860s, though, only a small number of routes accounted for nearly all the traffic from each. The northern or Wyoming field's Delaware and Hudson Canal shipped 1.64 million tons annually; the Delaware, Lackawanna, and Western Railroad carried 1.7 million tons; and the Pennsylvania Coal Company shipped 950,000 tons along its railroad line. The middle or Lehigh field split its traffic between the Lehigh Coal and Navigation Company's water route and the Lehigh Valley Railroad. Finally,

Gowen's Philadelphia and Reading Railroad carried nearly 3.6 million tons, and the Schuylkill Navigation Company brought a million tons to market via its canal. "It cannot be claimed that there is any combination between these carrying companies against the interests of consumers," New York's *Hunt's Merchants' Magazine* reported in 1869, "but the relations existing between the mining and carrying companies are so close and intimate that the results are practically the same." In 1873, Gowen met with the presidents of the anthracite railroads to form a formal cartel that would fix the percentage of traffic and maintain a stable price for anthracite coal of five dollars per ton in New York City. This price-fixing agreement, considered by many historians to be the first industry-wide arrangement in American history, allotted the PRR about 25 percent of the entire anthracite trade, far and away the largest share. The cartel created a board of control, which had the authority to inspect the books of each firm to ensure compliance, and for two years was able to stabilize the price of coal at or above five dollars per ton. Profits soared as a result. In 1875, for example, the PRR and Lehigh Valley Railroad paid their stockholders a 10 percent dividend, while the Lackawanna Railroad doubled its net earnings. Control of the anthracite trade did not, however, come without controversy. Coal dealers in Philadelphia railed against Gowen's combination, arguing in 1872 that "the natural laws of supply and demand should be the basis of trade" and that "healthy competition is of paramount importance in all business." Dealers were convinced that the overexpansion of collieries during the boom wartime years was the real cause of industry's instability and that Gowen's cartel placed an artificial floor on coal prices. "If the Reading Railroad President is so anxious to furnish consumers with cheap fuel," they asked by 1873, "why does he not reduce his freights?"[27]

The Workingmen's Benevolent Association joined the retail coal dealers in the fight against Gowen's cartel by 1874. The Schuylkill Coal Exchange, which formed under Gowen's leadership to coordinate the activities of colliers against the WBA, announced that wages for contract miners would be cut by 20 percent that year. The WBA had no choice but to call a strike, during which miners and mine laborers spent the first half of 1875 protesting the Schuylkill Coal Exchange's position. The PRR used its Coal and Iron Police, the private police force sanctioned by Pennsylvania's state legislature in the wake of Civil War–era conflicts in the anthracite region, along with private detectives from the Pinkerton Agency to provoke strikers at every turn. The WBA's financial resources could not sustain the prolonged struggle, and as it proved incapable of maintaining order, sporadic violence broke out across the anthracite region. Gowen and the Schuylkill Coal Exchange's plan to goad individual workers into pushing back against their hired police worked perfectly, and public sentiment turned against miners. The "Long Strike of 1875" ended when anthracite miners and laborers returned to work in June at wages that were 26.5 percent lower on average than 1869 levels. One labor organizer recalled that "evil days had come." "We went to work," he remembered, "but with iron in our souls." Gowen and the PRR reigned supreme after the Long Strike, leaving a private entity controlling a 250-square-mile coal-mining region with impunity.[28]

When violence continued to plague the anthracite region in 1875 and 1876, Gowen's campaign against labor spread to include a shadowy organization of Irish laborers in the anthracite region known as the "Molly Maguires." In Ireland's agrarian protests of the 1840s, the Molly Maguires became infamous for attacks and threats against landlords across the northwestern Irish countryside. By the 1860s, some Irish immigrants appropriated the name "Molly Maguires" as an implicit threat to mine operators in the anthracite region. Violence was sporadic and not endemic to the region's large Irish population, yet Gowen and other colliers sought to make it appear so. By hiring a Pinkerton detective, James McParlan, to infiltrate the Mollies and uncover their supposed terrorist agenda, Gowen hoped to wipe the anthracite region of labor militarism once and for all. To do so, he needed to blur the distinction between a respectable Irish American organization, the Ancient Order of Hiberians, and the Mollies. After a series of trials from 1876 to 1878, in which the PRR's private agents arrested men provoked into action and then accused by McParlan of violent crimes, twenty men were hanged. The charges against the Mollies pushed the limits of believability by portraying them as rabid terrorists who had erected a massive conspiracy designed to undermine the coal trade. And yet that legal team, headed by none other than Franklin Gowen, succeeded in convincing a carefully selected jury that this was the case.[29]

But just when monopoly control of the Pennsylvania's anthracite trade seemed within its grasp, the PRR's bid to control the flow of the nation's preeminent home heating fuel broke down. Gowen broke the WBA in 1875 and successful smeared the anthracite region's Irish workers with the Molly Maguire trials in 1877, but he also saw his cartel collapse at the same time. As it turned out, the incentive to cheat the cartel was too high for members of the association. When the price of anthracite nearly doubled to $5.28 a ton in 1875, the informal pool fell apart, the formal organization dissolved in 1876, and by 1877 the renewed competition for market share drove prices down to $2.34 a ton. Various alliances formed and reformed during the 1880s, usually under informal "gentlemen's agreements" among the railroads' management. But these eventually fell apart as well. As the region's capitalists failed to organize in Pennsylvania's stone coal fields, so did labor. Strikes and lockouts continued to characterize the industry, but of the seventy-three strikes among anthracite workers that occurred between 1881 and 1887, only sixteen were actually called by a union, and less than half of them were successful in improving wages or working conditions. Haunted by the high debt and perhaps the high personal cost of keeping the PRR in the public spotlight, Franklin Gowen's status and influence waned by the 1880s.[30]

His long battles with rival anthracite carriers, state legislators, coal dealers, and miners apparently took a heavy toll, as Franklin Gowen quite suddenly shot himself in his Philadelphia hotel room in 1889. Gowen's predecessor at the PRR, Charles E. Smith, thought that he ended his life "purely by mortification over his failure in the management of the road." When the post-Gowen PRR tried once again to control the anthracite trade in 1892, it failed and fell into bankruptcy. The sharp competition among coal carriers would not be blunted until J. P. Morgan reorganized the anthracite railroads using interlocking corporate directories in the early 1900s. Perhaps it is not surprising that Gowen failed in his attempt to corner the anthracite trade. As generations of colliers, carriers, and dealers knew, the American coal trade defied mastery and wrecked many would-be monopolists.[31]

The struggle among dealers, Gowen, and the WBA is only one short chapter in a long history of conflict within the American coal trade. The lesson in the complexity of home heating fuel markets, however, was repeated across the United States. The industrial household confronted a system with so many moving parts—dealers, wholesalers, carriers, miners, and colliers—that wrangling some sense of order from it seemed impossible. Even with the arrival of large corporations like the Philadelphia and Reading, which had a hand in nearly every phase of the trade, the flow of coal to American cities still eluded basic coordination. If Franklin Gowen and the PRR could not harness this massive system during the 1870s, then no single entity could; the river of coal continued to flow into urban markets, albeit with sporadic famines and gluts that so outraged its consumers and users. Industrial-era Americans had become accustomed to a steady supply of heating fuel in their cities, even as they complained about the dirt and grime involved in cleaning up after coal fires, the duplicitous behavior of coal dealers, the grasping ambitions of railroad executives, and the propensity of miners to strike for higher wages and better working conditions.

In many ways, the sheer size and complexity of the coal distribution network by the late nineteenth century afforded many urban Americans the luxury of remaining ignorant of how coal arrived to their hearth. "While seated at comfortable fire, how seldom, if ever, does one's thoughts turn to the condition of the miner who toils amid so many dangers to procure this comfort for us," mused one Boston coal dealer in 1869. In 1887 Charles Miesse scolded the readers of his *Points on Coal and the Coal Business* for being unaware of the high cost of securing cheap heating fuel. "So, my gentle reader, think while you are enjoying the blessings of coal, and growling about the high price of it, and heaping your anathemas on the coal men," he wrote. "They are men, brave enough to rush into the jaws of death, risk their lives, in the deep caverns of the earth, eat their bread in stench poisonous caves, and damp sickly dungeons, and waste their health, down in a dark sulphurous pit, away from the genial sunlight." "Complain to your plumber, not your miner, for the latter sacrifice a large percentage of life and blood, for your comfort," he concluded. "The small amount of money expended for coal, is a very trifling proportion of your yearly household expenses, and nothing gives you as much for so small an outlay." In the 1880s Terence Powderly, the leader of the Knights of Labor and three-time mayor of Scranton, Pennsylvania, asked a distinguished visitor to his home town, "Were you ever in hell?" The visitor, Father Thomas Ducey of New York City, toured the area's coal mines with Powderly, seeing firsthand the noise, confusion, and horrendous conditions in which nineteenth-century anthracite miners worked. "That is hell, sure enough," Father Ducey concluded. "I shall never complain of the price I pay for coal again."[32]

Urban Americans thus encountered an entirely new set of problems in heating the industrial household. Was cheap heat worth the trouble? Was there some other way to provide heat without the physical problems of dirt, ash, and ill health or the market disruptions caused by the constant fighting between miners, colliers, carriers, and dealers? By the 1870s, one potential way to replace the coal-burning stoves and grates appeared with the advent of district steam heating. This brand-new system involved a relocation of combustion from the home to a central factory. Could this be the logical outcome of an industrial society that valued standardization, centralization, and moving into larger and larger economies of scale? The application of steam, as the foremost power source of the industrial economy, appeared to be a logical solution to the ad hoc systems that had emerged to provide cheap stoves and mineral fuel to urban households. Yet making further alterations in the hearts of industrial America would not be so simple. As the next chapter demonstrates, the advocates of steam heat learned firsthand the difficulties of displacing the coal-burning stove and implementing an even more complicated system of home heating.

5 How Steam Heat Found Its Limits

When he arrived in Boston seeking his fortune in 1829, twenty-year-old James Jones Walworth struggled to find it. After three months of wage labor on a nearby farm, James entered the firewood-carrying trade between Boston and nearby Cambridge as a boatman. He found this work both sporadic and physically challenging; falling on a load of firewood one day, he painfully wrenched his ankle, putting him out of action for a few days. Even when fit, James complained in his diary that the boat's owner, Mr. Hills, "had nothing to do & would neither find me employ nor pay me for my work," as the carrying trade stalled in the late summer of 1829. Without the voracious winter demand for firewood, fewer and fewer boats were needed to transport fuel along the Charles River. James pawned his watch, did odd jobs like carrying local drunks to jail on handcarts, and pleaded with Mr. Hills for his back pay. "My prospects are yet dark & gloomy," he wrote in his diary on August 31, 1829. Walworth's experiences were not atypical for boatmen working to supply urban markets with energy; the firewood business had always been subject to gluts and famines. Eventually James abandoned the notion that he could make his fortune in home heating fuel and found work at a local hardware store.[1]

Even though Walworth gladly left the firewood trade behind, he remained active in the home heating business. His brother-in-law, Joseph Nason, developed an American version of the British "Perkins" system that distributed hot water through a network of pipes to heat large spaces efficiently and without the incumbent smoke and soot of closed stoves or hot-air furnaces. Nason and Walworth eventually formed a business partnership for installing these heating systems. After outfitting Boston's Eastern Exchange Hotel with a steam heating "apparatus" in 1845, Walworth & Nason received the $6,750.00 contract to develop a high-pressure steam heating system in the Boston Custom House. Walworth turned to other large-scale buildings, using either hot water or steam to replace individual fireplaces, stoves, or hot-air furnaces. "Our main business this past season," he wrote in 1846, "has been the erection of apparatus for warming factories, the method of which was originated about two years since of using small wrought iron tubes for heating by steam is becoming very popular." Walworth & Nason became pioneers in this field by practically inventing an entirely new heating system for their customers. They tinkered with new valves, elbow joints, fittings, and pipes of all sizes and shapes. To enable steam to heat a room efficiently, for example, James Nason developed a device of coiled pipes equipped with valves for controlling them. These "radiators" became a signature feature of steam heating and remained a part of James Walworth's systems even after Nason left the partnership. By 1853, James Walworth & Company had earned such a reputation for quality that he was invited to Washington, DC, to discuss a new heating and ventilation system in the executive mansion. With its tall ceilings and drafty hallways, the White House apparently was sometimes less than comfortable. Referring to one particularly frosty area, President Andrew Jackson reportedly had joked that "Hell itself couldn't heat that corner." Walworth couldn't help but agree when he visited. "Found the President's house wanting most of the improvements & conveniences found in Gentlemen's houses now in New England," he dryly noted in his diary.[2]

James Jones Walworth sold his customers on more than boilers and radiators; he sold them an entirely new system of home heating. In place of a number of individual stoves or fireplace grates to tend to, Walworth provided his customers an integrated and extensive network of pipes to circulate hot water or steam throughout the house. As the heated water or steam pushed through this network, it swirled around in the coils of each radiator and thus warmed each room. Once it cooled and condensed, the water traveled through a separate series of return pipes that would bring it back to a basement boiler for reheating. Walworth's systems applied the cutting-edge power of the Industrial Revolution, steam, to the age-old problem of heating spaces cheaply and efficiently. Applied to water and land transport to power steamships and locomotives, steam helped shrink time and space. When steam engines lifted water from coal mines, liberated textile mills from waterwheels and turbines, or shot air into blast furnaces, the transformative power of steam became apparent. Generations of historians have marveled over the monumental changes in industrial production that the application of steam facilitated over a few short decades in the nineteenth century. Could James Jones Walworth's system move from large-scale industrial buildings or palatial homes like the White House into the more modest homes of American cities? Walworth himself experienced the high human cost of the feast-or-famine arrangement that burning firewood entailed in the years before the rise of coal. Could steam heating systems offer remedies to the old problems of supply and demand? Was the application of steam to home heating part of the inevitable progress of industrialism?

Steam and Home Heating

The first use of steam and hot-water heating in the United States came in the 1840s, with the arrival of Walworth & Nason's system for warming both public buildings and textile factories and the construction of a high-pressure system to warm inmates at Philadelphia's Eastern State Penitentiary. In addition to systems integrated into the original construction of large buildings, contractors could install central heating in existing structures. Such was the case with the US Capitol, in which experimental methods of hot air and steam heating began as early as 1803 and continued through the 1850s. In general, large public and industrial structures were the most likely candidates for experimenting with innovations in hot-air or steam heating before the Civil War, often with the assumption that the results of these experiments would eventually applied to individual households. When a special committee of New York legislators considered the installation of a new $5,000 system of central heating and ventilation in Albany's State Capitol, they argued that "they may and ought, as servants of the people, not only protect and promote their own interests in matters of health and comfort, but in their wisdom devise and adopt the best conceivable plan for ventilating and heating this chamber." Doing so, the committee's 1847 report stated, "may serve as a model worthy of installation throughout the State."[3]

The transfer of hot water and steam heating from large institutional settings into wider home heating markets was a difficult proposition. Antebellum dealers in steam heat employed the same tactics as those used by hot-air furnace merchants; they focused on the health hazards and unpleasant side effects of using stoves or fireplace grates to heat rooms. The benefits of using steam instead of hot air, they argued, were even greater. Charles Davenport pitched the heat coming from his low-pressure steam heating systems as being "of the mildest and most agreeable nature" and assured that customers would be "relieved from all the dust and ashes" of stoves and fireplaces as well as the "dust that is driven through the register, when the rooms are heated by a hot air furnace." A Boston dealer in low-pressure steam heating furnaces, William G. Pike & Company, argued in 1857 that its "safe and economical" system offered a clear alternative to the "dry and uncomfortable atmosphere of the Hot Air Furnace." Stove maker Stephen Gold secured four patents for a steam heating system in the mid-1850s and enlisted the Yale professor of chemistry Benjamin Silliman Jr. to organize the Connecticut Steam Heating Company of New Haven. Gold then sold his patents, and Silliman's scientific celebrity endorsement, to a number of firms. The Massachusetts Steam Heating Company, for example, touted Silliman's approval of steam heat and reproduced large portions of the professor's article entitled "The General Principles of Artificial Warming and of the System of Mr. Gold in Particular," in which he denounced "the brittleness of the finger-nails, the dryness of the skin, producing an intolerable itching, and an oppressive sense of fullness about the head" caused by hot-air furnaces. The "General Principles" went on to describe steam's special heating properties, as it unleashed the power of "latent" heat in water rather than simply scorching the air. A dealer in Gold's patented system, the New-York Steam Heating Company, used the testimony of a physician, who found steam heat "exceedingly agreeable, having the softness of mild summer air, free from dust and dryness, and the escape of gas." Another testimonial, from Dr. Edward Bayard, noted that the Gold system had "so completely domesticated steam" that it actually had restorative properties for the human lung.[4]

Like the marketers of hot-air furnaces, though, antebellum steam heating contractors targeted a wealthy clientele with the luxury of experimenting with their method of home heating. "The introduction of hot-water apparatus was a worthy step from the wretched imposition of the quack stove-dealers, who might be aptly termed public poisoners," one engineering trade journal argued in 1853. And yet, the editors concluded, "this kind of furnace is too expensive" and "can only be afforded by the affluent." William G. Pike

& Company offered a testimony from a Waltham, Massachusetts, customer who found the "management of the fire so simple, that with the ordinary help of a house I have been able to keep it in successful operation." Dealers admitted the great expense of their systems. "We cannot furnish the apparatus at a first cost less than some other methods of heating," the Massachusetts Steam Heating Company admitted, "though we do claim a very decided superiority in this particular over any other *steam* or *hot-water* arrangement." In 1857 Henry Ward Beecher, the famous minister, described his struggle with the hot-water heating system in his new "capacious brown stone dwelling" in Brooklyn. At first, he marveled at the "complete arterial system" in his house: "the boiler being the heart, the water the blood, the pipes at the hot end the arteries and the return pipes at the cool end the veins." But when his daughters stoked the furnace one chilly day, they panicked at the rumbling noises of the pipes and doused the fire so quickly that they cracked the boiler. After carefully contemplating different patterns and schemes, the Beechers decided to enlist Richardson & Boynton to reconstruct a hot water system. After six weeks of work with "masons, tenders, ironmen, old iron and new iron, tin pipes, carpenters," Beecher reported that "the furnace took charge of the house." Although the system was expensive and time-consuming to construct, his final verdict on hot-water heating was positive. "A winter and a half on Brooklyn Heights will put any furnace to proof," he reported. "They may heap winter as high as they please without, we have a summer within."[5]

Dealers continued to pitch steam heat as healthy and cost-efficient alternative to stoves and hot-air furnaces following the Civil War. As in the antebellum stove trade, dealers stressed new innovations—either with or without a patent claim—that made their particular designs stand out from the crowd. Boston's Union Warming and Ventilating Company, for example, offered a self-regulating system that distributed steam heat without using radiators, thus adding to the "softness" of the heat. Even the most careful and skilled servant, its catalog argued, could not "hope to excel the quickness and delicacy with which this automaton answers every call for *more* or *less* heat." Baker, Smith & Company offered a series of weights marked "For Moderate Weather," "For Cold Weather," and "For Very Cold Weather" that could be pulled to open and close flues and thus regulate the temperature of a house. This New York firm offered testimonials from customers who attested to using only twelve or thirteen tons of coal per winter while keeping their house at an even seventy to seventy-five degrees. Such copious amounts of fuel suggest that Baker, Smith & Company catered mainly to the wealthy consumer, who in turn relied upon hired operators of the system to maintain it. The firm also denounced the absurdity of lavishing thousands on "luxuries and superfluities, while the air in our dwellings is poisoned and burnt by heating arrangements whose only recommendation is that they are *cheap.*" In 1872 an owner of Stephen Gold's patented steam heater condemned stoves and furnaces that "vomit[ed] forth a sirocco of *burned* vitiated air" and accused their dealers of purposely poisoning their customers. "People don't care what *kind* of heat you give 'em, only give it to 'em hot enough, and enough of it," one anonymous dealer is said to have bragged, "and they will never know the difference."[6]

Despite the aggressive claims of dealers, steam heating systems were not at all common in American homes. As in the very early stove trade, a limited industrial capacity was part of the problem. The first generation of pipes and boilers for home heating employed the same design as those for steam engines, which meant that foundries used an expensive form of wrought iron. The eminent stove maker Jordan Mott estimated in 1870 that 250,000 to 350,000 feet of radiator pipe had been cast in the entire United States, with only eight firms manufacturing boilers for home heating. Compared to the 275 stove manufactories, consuming 275,000 tons of iron and assembling 2.1 million stoves, the manufacture of steam heating components was still in its infancy in 1870. Eventually more cost-effective cast-iron boilers and radiators appeared on the market, and by 1880 the number of foundries making boilers for heating had increased to eighteen and radiator pipe to a little under two million total feet. This is an impressive increase, but it was not enough to bring costs down to a level competitive with stoves and hot-air furnaces. The potential customer base put another limitation on steam heating sales. By the early 1870s, leading firms like James Walworth & Company and the H. B. Smith Company dealt mainly with large-scale institutional structures. The latter laid claim to four hundred boilers installed in individual homes since its founding in 1860, but its patented Gold cast-iron boilers were reserved almost entirely for the wealthy. Walworth's customers also came from the ranks of the elite. In 1859 he contracted with Illinois governor John Wood for a $4,500 heating system in his Quincy mansion and also met with Chicago railroad baron William Ogden about heating his house with hot water. These were both custom-designed systems that offered very little inspiration or potential for wider markets.[7]

While limited industrial capacity and customer base were the primary impediments to the spread of steam heating systems, perceived fire risks created another barrier to their popular adoption. One Chicago insurance underwriter refuted the claim that steam heat was safer than stoves or hot-air furnaces. Because "men usually in charge of such boilers are not experienced engineers, but often very ignorant of the proper management of steam," Arthur Ducat argued in his guide *The Practice of Fire Underwriting*, "in our opinion, [steam heat] has been the cause of many very disastrous fires, that have not been accounted for in any other way." In 1873 the *Boston Journal of Chemistry* lamented the tendency of water in steam heating pipes to freeze and leak out onto carpets and into walls. The editors further asserted that the complexity involved in managing steam heating systems was beyond the capacity of most users. "A large proportion of housekeepers have not the requisite skill to manage a parlor stove properly," they argued, "and if a sly, slippery fellow like steam is put in their charge, he is master at once, and sure to give her servants infinite annoyance." Some engineers cited the tendency of pipes installed in close proximity to woodwork to dry it out, much like, in the words of another 1873 assessment, "the process of kiln-drying, prolonged throughout the years." By 1877, another underwriter suggested that insurance companies void their policies on homes that allowed steam heating pipes to come into contact with woodwork. Although the explosion of a steam boiler might seem a more spectacular risk, the threat of fire was the most common complaint among skeptics of steam heating's safety.[8]

Along with fire hazards, the supposed health benefits of steam-heated air became a point of debate. Dealers and contractors avowed that steam provided "gentle and soothing" warmth much superior to the dry and hazardous atmosphere created by fireplaces, stoves, and hot-air furnaces. Benjamin Silliman, the scientific collaborator on Gold's patented system, argued that steam-heated rooms were "generally considered warm enough" at 65 or 68 degrees, while hot-air furnaces often drove the temperature up to 70 or 80 degrees. The reason, Silliman and other steam heating proponents argued, was that the latent heat of steam became "sensible heat again only when the steam is re-converted into water." *Scientific American* denounced this claim in 1868 as a "scientific absurdity" and alleged that if Silliman meant that "air is more moist when heated to the same degree by steam than when heated by hot-air furnaces, an error is committed." Instead, the editors argued, proper ventilation and the use of small steam jets to maintain a room's humidity should be employed. In fact, the tendency of steam heat to create lower temperatures in rooms, rather than an innate quality of its latent heat, was the real reason that steam-heated air seemed less dry and irritating. When university leaders contemplated the installation of central heating in student dormitories in 1869, one editorialist in the *Harvard Advocate* found the atmosphere of steam heating to be "very disagreeable, if not injurious" and noted that "few persons [were] able to stay even a short time in a room thus heated without contracting severe head-aches, eye-aches, and many other discomforts caused by the oppressive quality of this heat." Pneumonia, one medical journal claimed, was a specific disease brought on by steam heat and its "hot, bad air," which had been "long ago exhausted of its vitality."[9]

Steam heat thus ran into several layers of resistance to its widespread adoption in the 1860s and 1870s. Although it seemed suitable for use in office buildings, greenhouses, public buildings, and hotels, the high expense and complicated installation soured most consumers on adopting steam heat in their own homes. As a result, central heating remained the province of wealthy households through the World War I era. Central heating systems appeared beyond the reach of most American consumers as long as the financial barriers remained in place. The implementation of steam, hot water, or hot-air furnaces in American cities could not follow the same process as the adoption of the coal-burning stove. The price tag remained too high, and central heating's move from affluent to modest households was slow. Apartment buildings that catered to an exclusive, wealthy clientele that boasted all the modern amenities of running hot water, individual bathrooms, and gas lighting added steam heat to their lists of luxury comforts. But the installation of individual steam heating systems in individual households across the economic spectrum simply was not economically feasible in the cities of industrial America, as the minimal demand generated among consumers seemed unlikely to support a growing, diverse, and innovative trade like the one spurred by the market for coal-burning stoves a generation earlier.[10]

Home Heating and the Networked City

If consumer demand lagged, though, perhaps companies could sell steam heat directly rather than producing it within the home, via a new kind of "networked technology." However, the explosive and haphazard growth of American cities in the half century after the Civil War already overtaxed the rather modest facilities in transportation, water, sewage, and communications in place. Over the course of decades, cities witnessed the growth of an infrastructure of pipes bringing water and natural gas into urban households, along with sewer lines carrying away waste materials from those same homes. Streetcars and paved streets replaced the haphazard traffic patterns of antebellum cities. In the 1880s innovative "networked technologies" such as electrical power and telephony offered new forms of energy and communication within this urban infrastructure. By the advent of the twentieth century, then, several overlapping networks had revolutionized the way that Americans traveled to work, lit and cleaned their homes, and talked to each other. But the development of networked technologies always occurred within a political and economic institutional framework that could both encourage and resist change simultaneously. If the advantages of steam heat could offer a new networked technology for home heating markets of the late nineteenth century, it wasn't likely to unfold in a logical or systematic fashion. Nor would it appear overnight.[11]

Home heating markets of the late nineteenth century were, of course, already working within a national network of sorts. In the 1850s and 1860s, as we have seen, delivering coal to urban households involved miners, colliers, railroad and canal companies, wholesalers, and coal dealers in a loosely integrated system in which several actors worked in their own interest. At various points these networks broke down, and consumers and users certainly voiced their displeasure with individual components—miners when they went on strike, railroad executives when they tried to corner anthracite markets, coal dealers when they gouged their customers or delivered a short load—and yet because this was only a loosely integrated system, it was impossible to rework it completely. In other words, the power to change the mineral fuel network was distributed among so many actors—consumers, dealers, state legislatures, corporations, unions—that a significant overhaul of it was practically impossible. At the same time, however, the growing size and power of one of these vital links in the mineral fuel commodity chain, the railroads, offered an example of how large-scale organization might make sense of this cacophonous mix of market transactions. The managers of railroads needed to coordinate the transport of goods and people over long distances, adhere to strict timetables, and turn a profit. In order to meet those challenges, they created large-scale hierarchical frameworks that facilitated the flow of information from sales agents, accountants, engineers, and workers up to managers, who then sent directives back down the lines of communication. Theoretically, railroads acted with military precision in coordinating America's first "big business." In reality, these first movers of corporate integration encountered snags and backlogs quite frequently; anyone familiar with "military efficiency" knows the gap between strategic plans and the execution of those orders in the field. Nonetheless, the organizational innovations in American railroads provided a kind of blueprint that allowed for a tighter integration of several business functions once left to the free market. Since nearly everyone involved in the process had some complaint about the way in which mineral fuel was produced, transported, and distributed to consumers and users in American cities, perhaps the rise of these new methods of integration offered a model for revolutionizing home heating.[12]

In fact, there are many historical examples of networked technologies that integrated production, distribution, and consumption of energy during the Industrial Revolution. The emergence of networks for selling illuminating gas, for example, dates back to London's Gas Light and Coke Company (GLCC), founded in 1812. The GLCC controlled every aspect of the gas industry—it built a gasworks to convert coal into gas, constructed a series of pipelines to transport it, and contracted with fitters who installed lamps in their customers' homes. The GLCC also struggled with complaints from consumers and users, who wanted to light their homes around the clock rather than adhere to the company's distribution schedule. By adapting its production and distribution organizations to meet these challenges, the GLCC offered one example of a tightly integrated system for delivering gas. Another example of a networked technology in energy supply is the rise of electric power systems during the late nineteenth century, exemplified by Thomas Edison's combination of invention and entrepreneurship in constructing his early direct-current electrical power network. The spread of electric power eventually blended public and private interests, albeit imperfectly, with the rise of public utilities. In doing so, urban electrical networks coordinated large infusions of capital and technological innovation in providing an entirely new type of commodity to urban consumers. Eventually, a larger and more complex version of electric power service pioneered by early innovators like Edison in large markets such as New York City spread across the nation to revolutionize everyday life in both urban and rural settings.[13]

If the encroachment of large technological systems into the American home seemed a foregone conclusion by the late nineteenth century, large corporations or public utilities were essential to this process. A recent study of the nation's telegraph and telephone networks describes stages of "commercialization," in which public and private actors develop the system; "popularization," in which these communication networks served mass audiences; and finally, "naturalization," as when the telegraph and telephone shed their political and cultural novelty and became integrated into the normal course of life. As more sophisticated systems like telephony connected more and more homes into large-scale technological networks, households became consumers of new commodities that seemed at once indispensable, tightly integrated, and yet also ethereal. How does one "consume" light, power, a conversation, or heat? Over the course of the nineteenth century, the use of coal-burning stoves or grates in fireplaces had become integrated in most urban homes, and the network of delivering mineral fuel to the urban North had been commercialized, naturalized, and popularized. And yet the delivery of mineral fuel was by no means perfect. Customers still needed to bargain with coal dealers, who complained about being gouged by railroads and coal companies, who in turn griped about striking miners and their potential to interrupt the flow of coal. Instead of purchasing, shoveling, and burning coal, wouldn't flipping a switch or turning a dial in the comfort of your own house be an easier way to stay warm? Could the advocates of steam heat replace home heating fuel's loosely organized and chaotic system with one that used large, organized, and well-financed corporations or utilities that could deliver a service?[14]

The engineer Birdsall Holly certainly thought so. In the 1870s, Holly built a prototype boiler and pipe system to heat several square blocks in his hometown of Lockport, New York. His idea was to centralize steam generation and then distribute it throughout an entire community, in essence expanding the scale of central heating from the individual home to a citywide network. This was not a wholly new concept. Arthur Brisbane, the antebellum social reformer, denounced the "ignorance, cupidity, carelessness or inability of individuals" and advocated centralizing heating functions in his design for collectivist communities called "phalanxes." Brisbane sought to unlock the productive capacity of humanity through the mathematical principles of Charles Fourier in 1840; Holly had less philanthropic but equally ambitious ideas about rationalizing the way that Americans heated their homes three decades later. Holly's "district heating" system required several innovations in the design and construction of mainline pipes, meters for measuring usage, and distribution pipes for spreading the steam throughout Lockport's single-family households. The pipes that delivered steam directly to homes were insulated with "mineral wool," or asbestos, and Holly designed a series of regulators and expansion joints that kept the pressure of the system at a safe level. Within the homes, condensation traps gathered water, with the assumption that families could use it in their everyday chores. In fact, Holly hoped that new methods of cooking with steam would catch on and make the system even more valuable for domestic markets. Herman Haupt, a prominent engineer and early advocate of railroads getting into the coal trade, visited Lockport and wrote a report on the Holly system in 1879. Haupt concluded that Holly's "atmospheric radiators" heated rooms efficiently, safely, and without any of the storage or cleaning issues of the coal-burning stove: "Pipes are painted and bronzed, and are quite ornamental; the space occupied and the cost are trifling, all making of fires and attendance are dispensed with, and the annual cost of fuel greatly reduced." He estimated that district heating cut the cost of heating a room by more than 70 percent and that Holly's invention stood as an agent "of the highest order in advancing the interests and promoting the comfort and happiness of suffering humanity." "I may not live to see the bright dreams of the future realized," Haupt mused, "but if capitalists will refrain from excessive charges, humanity will reap large benefits from their inventions." But along with his mention of the Holly system's civilizing influence, Haupt also calculated that investors could reap dividends of up to 66 percent on their investment.[15]

Haupt's breathless endorsement of the Holly system should come as no surprise, as he was a paid consultant charged with assessing the feasibility of the network to larger cities like New York. In anticipation of this expansion, Holly patented his innovation and formed the Holly Steam Combination Company. In essence, Holly franchised district heating and offered to design and supply separate companies formed in individual cities, thus providing both an efficient heating method and an avenue to profit. It was a privatized solution to the challenge of networking the technology of home heating. By 1881 fifteen commercial steam companies were operating in cities as diverse as Springfield, Massachusetts; Detroit, Michigan; and Denver, Colorado. There were problems; many early adapters didn't construct their systems to Holly's exact specifications and lost a great deal of steam to condensation. In other cases, Holly's meters did not work, and district heating companies charged customers a rate based on their prior coal bill. Despite early setbacks, the Holly system continued to find backers who wanted to adapt this system to America's largest cities.[16]

Although it was not the first city to adopt the Holly system, New York City provided the highest-profile experiment to determine the benefits of district heating. The project began with Wallace A. Andrews, one of the original directors of the Standard Oil Company, who had visited Lockport and was so impressed that he hired an engineer, Charles Edward Emery, to bring the system to New York City. In 1879, Andrews acquired the exclusive right to implement the Holly system and formed the Steam Heating and Power Company of New York. Andrews merged his company with a rival to form the New York Steam Company, and in 1881 the company began to dig up the streets to lay pipes—often at the same time that Thomas Edison was laying electrical mains for his direct-current power plant on Pearl Street in lower Manhattan. The New York Steam Company had a main plant on a plot bounded by Cortlandt, Dey, Greenwich, and Washington Streets with forty-eight 250-horsepower boilers, sixteen on each of the three floors, with a 225-foot chimney—the second-largest structure on Lower Manhattan. In 1882 the company began serving its first customers, drawn mostly from the immediate area of the firm's main plant. Expansion of the steam network would soon follow. The editors of the *Sanitary Engineer* called the New York Steam Company's plan to construct a system with twelve to fourteen stations throughout the city "the most stupendous undertaking ever contemplated for the production of steam, that has a prospect of success" and lauded the company's "practical demonstration" of district heating with steam. Like Edison with his Pearl Street electricity station, Andrews and Emery sought to reorganize the way that New Yorkers purchased energy.[17]

The implementation of district heating in New York City drew a great deal of attention—both positive and negative. *Scientific American* highlighted the project in November 1881 and announced rather breathlessly that steam heat delivered through underground pipes would "banish fire completely from our dwellings, office, and factories." The editors explicitly compared the project to Edison's attempt to electrify lower Manhattan. Other voices, however, found the New York Steam Company's pipe laying a major annoyance. Eventually the Department of Public Works fielded requests to limit steam-pipe laying to the evening hours. The editors of *Puck* articulated discontent quite fiercely and with characteristic sharp wit that same year:

> It will probably be conceded that the people of New York have some rights in their streets—not many, of course, but one or two. One of these must be the right to use those streets for the purposes of business. If this assumption be not too strong, is the Steam Heating and Power Company to make those streets impassable whenever it sees fit, simply because its incompetent engineers were so insanely foolish as to put down pipes incapable of standing the pressure upon them? Is there to be no limit upon this thing?[18]

The laying of the steam mains seemed bad enough, but problems continued to plague New York's district heating network after their completion. When a customer ordered service, for example, a trench had to be dug for the feeder pipe. When leaks in the pipe network developed—a frequent occurrence according to anecdotal sources—"clouds of offensive vapor" shrouded streets and nearby basements. One reporter for the *New York Tribune* counted ten obstructed streets in one small area of lower Manhattan in late 1882. Leaks could also lead to explosions. The Citizens' Steam Company's pipes in Lynn, Massachusetts, exploded three times in two weeks in the summer of 1882, sending stones and gravel fifty feet in the air and injuring several people. New Yorkers feared that a similar event on the Broadway steam mains "would be a serious matter, even if it amounted to only the hurling of a few hundred paving-stones over the passers-by and through the plate-glass show-windows which line the street." John Newton, the former commissioner of public works for the city, highlighted the problems of the steam lines in a report on New York's sewer system in 1888. Leaks in steam pipes, Newton argued, eroded underground electric lines and wore away at the cement holding together brick sewers, undermining both services and releasing "disagreeable odors" all along Broadway.[19]

Their pipe-laying program seemed to inconvenience New Yorkers of all stripes, but the New York Steam Company attempted to cultivate only wealthy consumers in order to highlight the benefits of district heating. Once they extended a line up Fifth Avenue, they counted New York's wealthy and powerful among their clients. The company drew testimonials that relied upon Andrews's Standard Oil connections: Henry Flagler reported that "the service has been entirely satisfactory," and John D. Rockefeller announced that "I have had my house heated for several seasons by steam supplied by your Company, and am satisfied with the service given." This ran counter to Haupt's 1879 appraisal of the system, which assumed that Holly's innovations would help poorer consumers. "The cost of fuel measured by steam consumed is almost incredibly small," Haupt had argued, "and for the poorer classes occupying tenement houses, the stove could serve as a radiator and furnish all the heat necessary, as well as cook the food, at a cost of a very few cents daily." Haupt concluded, "The combination of the steam stove with the Holly System would be a great boon to all classes, but to the poor especially." However, the early instructions for using the system described a fairly extensive system of heating coils, traps, and meters, which assumed that steam heat would be used in a multiroom dwelling (fig. 5.1). After describing how ladies would "learn to accustom themselves to this temperature in ordinary house dress," while "gentlemen" might need to change coats on entering the house, the circular promised that in due time, "the attendant will learn just what coils are required in different kinds of weather to maintain a uniform temperature." This strategy of highlighting steam as the "fuel of the fashionable" had parallels in the early adoption of anthracite coal, yet boosters in that case expanded their program to include less affluent consumers relatively early in the process.[20]

A massive blizzard in 1888 provided a brief highlight for district heating, as its clientele remained warm and toasty while the rest of the city suffered from the inability to deliver coal to households. Unfortunately, the blizzard also highlighted the affluence of the New York Steam Company's clientele and the failure of the system to benefit much of the city's population. The "Great White Hurricane" that socked the Northeast that season dropped as many as fifty inches of snow on New England and resulted in at least four hundred deaths. Fuel shortages ripped through poor communities. The *New York Times* reported that in the tenement houses of the Lower East Side, "women and children were running with pails from grocery to grocery in every part of the east side vainly trying to buy coal" and that in some cases "they had absolutely nothing in their tenements except their scanty furniture with which to maintain a fire." Although the New York Steam Company's main plant nearly ran out of coal, clients of the company enjoyed uninterrupted service—a development that the company placed at the center of its official history years later. But in addition to demonstrating the potential of district heating, the success of steam heat during the Blizzard of 1888 also revealed the limited reach of the company. Simply put, the New York Steam Company did little to widen its customer base to include less affluent consumers—a trend that became vividly evident when New York's poor shivered while the city's coal distribution network froze up during the storm.[21]

THE APPARATUS OF THE NEW YORK STEAM COMPANY.

DIRECTIONS FOR USING.

In the cut, *A* is the steam meter, and *B* the meter register. The series of pipes and valves at the right is called the meter combination. *S* is the street steam pipe, *P* the power pipe, and *H* the heating pipe. *C* is the service steam trap, *d* is the regulating valve. *Left hand* meter combinations have the valves similarly arranged but on the *left* of the meter. Occasionally regular meter combinations are not used, but valves can always be found corresponding to those shown in the cut.

To Turn Steam on and off the Meter Combination.

The inlet valve *b*, in the lower horizontal pipe is usually marked with a red tag, and should be always kept wide open, so as to insure a full supply of steam to the meter. If this be neglected, steam is supplied at reduced pressure, and is necessarily charged for at the higher pressure at which the consumer is entitled to have his steam measured. When steam is not required, the supply from the meter should be shut off by closing the *outlet* valve *c*. The wheel of this valve, or the one corresponding thereto, is usually marked with copper wire to distinguish it. When steam is shut off in this way, the meter is kept hot and steam may be turned on to the power pipe *P* again at any time by simply opening the outlet valve *c*. Before doing this a drip, near the engine or other apparatus supplied from the power pipe, should be opened to allow the air and water to escape.

To Turn on and off the Heating System.

Steam for heat will first be turned on by the employees of the Company and may be shut off when not required by closing first the outlet valve *h* of the regulating valva and next the inlet valve *g*, without interfering with the supply of steam for power. These valves should also be closed when the meter is shut off, as explained in the preceding paragraph.

To Start the Regulating Valve.

First open wide the outlet valve *c* of meter, then open the outlet valve *h* of the regulating valve a little, and afterward open the inlet valve *g* gradually, until it is wide open, when the outlet valve *h* should also be opened wide. Should the regulating valve jump while being heated up, a little steam may be first passed through it, and the pressure introduced to the heating pipe *H*, by the use of the regulating valve pass-by *i*. When everything is heated up, the regulating valve may be started as explained, and pass-by *i* closed.

To Start the House Trap.

There will be located on the returns from the heating system a house trap resembling the service trap *C*, on the cut of the meter and combination, but generally of larger size. The water from the house returns will be admitted through the inlet *n*, and discharged to the street return pipe through the outlet valve *p*. In general it will be sufficient to open the outlet valves *m* and *p* at the time of turning on steam. In connection however with the house trap, it is desirable to have a blow-off from the house returns into a sink or bucket, which can be opened on turning on steam, to hasten the circulation by letting out the air and water more promptly than it will pass through the trap. It is against the rules of this Company to lead this blow-off to the sewer, as it frequently becomes leaky and is occasionally left open, occasioning a very considerable waste which cannot readily be detected. The small valve *r* is used for testing the trap by the inspectors of the Company. Just back of it is shown a small air valve in the bonnet of the trap which may be opened to let the air out of the trap if the steam does not readily circulate so as to heat the metal. Special traps will be explained by the inspectors of the Company.

To Change the Heating Pressure.

The heating pressure may be increased by putting on more weights or decreased by taking off weights, each weight representing approximately one pound pressure. There must be sufficient weights kept on to maintain a pressure which will return the water of condensation to the street. This will require from four to eight pounds according to location of the premises. If a less pressure is required a special trap must be used which will be explained by the inspectors of the Company. When a spring is used instead of weights it will be better in general to have the adjustments of pressure made by the inspectors of the Company. In case the heating pressure varies so as to cause suspicion that the regulating valve has become stuc[k] from particles of dirt carried along with th[e] steam or otherwise, a slight jar on the yoke o[f] the valve may loosen the obstruction. I[f] not, the use of this valve should be discon[-]tinued by closing the valves *h* and *g*, an[d] the steam supply for heating be admitte[d] through the heating pass-by valve *i* to main[-]tain the desired pressure as shown on stea[m] gauge *f*. This will require attention unt[il] the Company is notified and the regulatin[g] valve put in order in due routine. Low pre[s-]sure heating apparatus should in all cases b[e] provided with a small safety valve which wi[ll] open and give notice to the attendant whe[n] the regulating valve becomes inoperative fo[r] any reason.

To Regulate the Heat in the Building.

The temperature of separate rooms may b[e] regulated by adjusting the registers, when a[n] indirect system is used, and by shutting o[ff] and turning on direct radiators when th[e] same are placed directly in the room. Th[e] temperature of the whole house may b[e] changed by varying the steam pressure, a[s] previously explained, being careful not to ru[n] it so low in any case that the water of con[-]densation will not be returned to the street. When building is still too warm, with lowes[t] pressure admissible, shut off progressively th[e] direct radiators and the registers of the in[-]direct radiators, and if necessary have som[e] of the coils and risers shut off in the base[-]ment. When the custom prevails of keepin[g] steam on all night it will be found unnecessary except in the severest weather, to use all o[f] the heating coils. After a little experienc[e] the attendant will learn just what coils ar[e] required in different kinds of weather t[o] maintain a uniform temperature.

It is recommended that thermometers b[e] placed in different parts of the house, and car[e] taken to maintain a temperature of sevent[y] degrees in all connecting rooms habituall[y] used by the family. Ladies, who frequentl[y] require a higher temperature, will in due tim[e] learn to accustom themselves to this temper[-]ature in ordinary house dress, and gentleme[n] desiring a lower temperature will, by changin[g] coats on entering the house, readily becom[e] satisfied with this compromise. It will b[e] found that the maintenance of a uniform tem[-]perature, not higher than that stated, wil[l] greatly aid in preventing all members of th[e] family from catching cold on passing from th[e] house to the street.

A comfortable moisture may be maintaine[d] by arranging for a slight escape of steam fro[m] one or more of the heating coils.

CHAS. E. EMERY,
Manager and Engineer.

General Offices of The New York Stea[m] Company, No. 22 Cortland Street, N. Y.

Up-town offices, 58th St., near Madiso[n] Avenue.

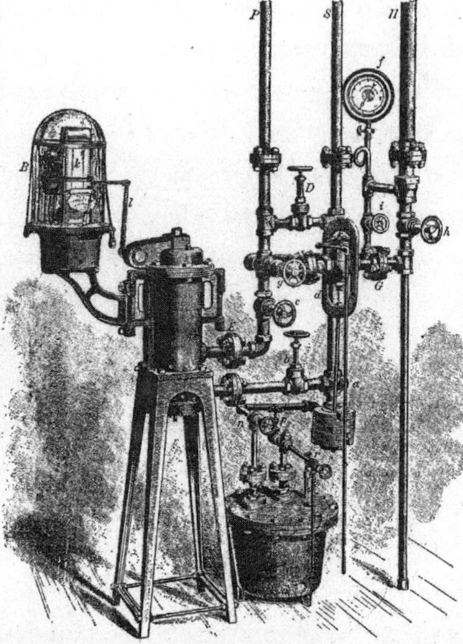

Figure 5.1. Instructions for the district heating customer. If district steam heat was to become a reality, customers needed a simpler way to use it. These instructions from the New York Steam Company demonstrate the complexity of its system. Most urban residents would have found the maze of pipes, gauges, and condensation traps rather difficult to negotiate. The lengthy textual explanation did little to ease fears that this was a highly intricate system. New York Steam Corporation, *Fifty Years of New York Steam Service: The Story of the Founding and Development of a Public Utility* (New York: New York Steam Corporation, 1932), 49

•

Actions by city officials in 1890 stunted any boost that the New York Steam Company might have experienced during the Blizzard of 1888, as both the Board of Health and the Board of Public Works slammed the firm for its impact on daily life in the city. The Board of Heath acted first by declaring the steam mains on Broadway a public nuisance. The threat of explosions, leaks that seeped into residential basements, and increased heat in the city's sewer lines all created "conditions dangerous and detrimental to life and health." A few days later, New York's commissioner of public works, Thomas F. Gilroy, declared that he "had no reliance whatever" in the New York Steam Company. The Board of Public Works revoked all of the permits allowing the company to work on its pipes and issued an order to arrest any employee of the company found opening city streets. When Andrews complained, Mayor Hugh Grant denounced the company's "reckless disregard of the rights of this municipality" and backed the order to shut off the steam mains. The editors of the *New York Times* took note of the company's elite clientele when it called for a revocation of the company's charter, even as the paper recognized the savings that steam heat offered its customers. "How much that economy amounts to in individual cases we do not know," the *Times* argued, "but if it is a benefit at all it is a benefit to a few persons, while the injury is borne by many who have no interest in the steam-heating business." The New York Steam Company reopened its steam mains after fixing some of their problems but soon thereafter shifted its strategic goals toward large institutional clients and never revolutionized home heating as it aspired to do only a decade earlier.[22]

New York City's experience with district heating offered a dramatic reminder of the system's drawbacks, but the more mundane problems of how to "consume" heat also undermined steam heating's feasibility in both large and smaller markets. In New England, for example, the Boston Heating Company used a system designed by Washington, DC, engineer William Prall, in which superheated water was turned into both high-pressure steam for power and low-pressure steam for heating. This company enjoyed the backing of Theodore Vail, the wealthy and influential president of American Telephone and Telegraph, who invested $215,000 of his personal fortune in the venture. The Boston Heating Company predicted that the popularity of providing "many of the prominent houses in our district with heat and power" would prove immediately profitable and claimed that as long as their customers avoided an "abundance of piping," steam heat was a cost-effective home heating solution.

Boston's city engineer regarded the placement of high-pressure steam lines as "dangerous to the public" and warned against the full implementation of the system, but Vail and his fellow investors pressed on. Once the system actually came into being, however, construction costs far exceeded estimates, and the Boston Heating Company was running low on money. Vail personally backed a cash infusion to the firm even as he confessed to a fellow stockholder that "this will take us out of the soup—or put *me* into it." In the end, though, the backing of prominent investors like Vail could not save the Boston Heating Company. The firm was dissolved in 1889—only a few years after its creation—as a result of ruptures and chronic rust developing all along its steam mains.[23]

The story of the Harrisburg Steam Heat & Power Company, which was formed in 1886, demonstrates some of the challenges of introducing consumers in a midsize city to district heating. After receiving a state charter that gave the company permission "to occupy all of the streets and avenues of the City of Harrisburg," it contracted with Lockport's American District Steam Company to construct steam mains based on the Holly system. "You have no care, you provide no fuel, you avoid the risk of fire, and yet you are always warm," the franchisers of district steam heating argued. "What other system contributes so much to the happiness and comfort of the people?" The apparent dangers and inconveniences of the system were a thing of the past; the American District Steam Company dismissed the "so-called explosions" and claimed that they were mainly the result of illuminating gas explosions coming from leaky pipes. Since steam escaping from pipes when ruptured caused a scalding no worse than that from a tea kettle, critics had no right to blame steam heat. Enough residents of Harrisburg were sold on these claims, and on March 21, 1887, the Harrisburg Steam Heat & Power Company began to supply steam to thirteen initial customers.[24]

Cultivating a customer base in this market required overcoming several challenges. First among these was measuring and billing for the heat used by each client. The Harrisburg Steam Heat & Power Company used Holly steam meters, which measured the steam as it entered each individual building. "The meters were fairly accurate unless tampered with," the company's historian recalled, "and in a number of cases this was done, by screwing up the adjusting screw and throwing the gears out of the mesh." Getting customers to rely upon meters to give an accurate reading of the steam used every billing period took time and effort from the company's directors. Some residents simply hired wildcat steamfitters to tap into the company's steam main. Others who were clients of the company complained bitterly about the accuracy of meters. "As was anticipated the meters have not given universal satisfaction, especially where the apparatus has not been constructed by your company," the company's president, E. Z. Wallower, reported in 1888. "The meter problem, like many others which have had to be encountered in a new business, is fast nearing a satisfactory solution." In essence, the company had to simultaneously convince its clients to use the company's own authorized steamfitters and equipment and then trust the accuracy of those meters at a time when district heating was still at a very experimental stage.[25]

The Harrisburg Steam Heat & Power Company found it difficult to extend the benefits of district heating to a wide base of consumers, as the cost of putting in steam heat was between $200 and $500, depending on the building. Customers paid for laying service pipes to the main at the rate of one dollar per foot from the curb to the inside of the cellar wall and also for the regulating valve, steam trap, stopcock, and other accessories. Although they paid a discounted rate for this equipment, the initial startup cost became a major obstacle. Even if the tenants of residential buildings wanted steam heat, President Wallower noted in 1889, the owners were often unwilling to pay for the installation, and the tenants refused to pay a surcharge of 10 percent of their rent for the use of steam heat. Initial start-up costs were therefore a problem for potential consumers that restricted the expansion of the system.[26]

Because some customers wanted to pay a one-time fee for their heat rather than receiving a regular bill based on their metered use of steam, in 1888 the company issued a circular that year to justify its billing practices and explain how steam heat should be consumed. "In answer to numerous requests that we contract for a season's supply of steam," the circular argued, "we beg to say that we cannot pursue such a course, and, at the same time, be equitable with our consumers." Its customers, furthermore, should understand that "a local Merchant could not contract to furnishing coal for a winter season regardless of the quantity consumed." "Heat, in any form, must be sold by measurement; if in the form of coal, it must be measured by weight; if in the form of steam, it must be measured by meter." By 1890, the company imposed a minimum and maximum charge to its clients, explaining that even if individual customers turned off their heat, the company still needed to keep steam waiting in its mains for other clients. An 1890 circular chided customers for their wasteful behavior and hoped that the maximum rate "will not permit the cooling of the premises by the raising of windows on the opening of doors, but contemplates the shutting off of steam when occasion requires."[27]

The Harrisburg Steam Heat & Power Company's problems seemed directly related to the novelty of the district heating plan. Harrisburg residents were used to burning wood or coal for heat, with a familiar rhythm of consumption. One bought coal for the winter season and then adjusted the rate of fuel consumption with the weather. On colder days, a stove or air furnace would use more coal; on cooler days it would use less or even shut down completely. Consumers could adjust the temperature of their dwelling and their rate of consumption on their own. Steam heat, on the other hand, required several leaps of faith. One had to trust the company to keep the pressure up in the steam mains, to provide reliable and accurate equipment, and to bill at a fair rate. The initial cost of installing steam heat was high and required using the company's licensed steamfitters, which further constrained consumers. Simply put, district heating asked quite a bit of late nineteenth-century customers. The company's clients expanded steadily in number in the first twenty years, but like its counterpart in New York City, the Harrisburg Steam Heat & Power Company failed to replace traditional forms of home heating in its market. By 1900 the company had 230 customers in a city with over 50,000 residents. The company's historian estimates that there were about eleven occupants on average per customer, which meant that the company served a blend of commercial and residential clients, with the former becoming more prominent as the company grew. In fact, its two institutional customers, Dickinson College in Carlisle and the Pennsylvania State Capitol in Harrisburg, provided a large source of revenue and accounted for a great deal of its steam production. So, even as its absolute customer base grew, its transformative effect on individual home heating preferences remained slight.[28]

District heating seemed an ideal way to bring steam heat into popular use. Its implementation, however, proved more difficult. In New York City, the many impositions that the maintenance of steam mains required on busy streets undermined the popularity of the company. The close proximity of steam to electrical, gas, and sewer lines created even more problems and eventually became a public nuisance. Boston's experience reinforces the notion that despite the backing of wealthy and influential investors, technical problems in implementing these systems were significant. Finally, the Harrisburg case suggests that consumers struggled with the high initial costs, service requirements, and billing methods required by district heating. Although district heating became an important part of commercial markets and, as it remains today, a popular choice for institutional clients, Haupt's vision for the Holly system failed to come to fruition. And although district heating would eventually play a large role in European housing and in the construction of large skyscrapers in the twentieth century, for which chimney stacks were unfeasible, it failed to replace the household hearth in the United States.[29]

As we have seen, earlier technological innovations in home heating followed a pattern of diffusion from wealthy consumers who have the luxury of experimenting with new methods of heating their residences and the ability to pay high start-up costs. Coal became Philadelphia's "fuel of the fashionable" in the 1820s, but the coal stove became an essential part of the working-class home only a few decades later. The two examples of district steam heating in New York City and Harrisburg demonstrate that this system may have reached the first stage of this transition—it was wildly popular with an elite set of customers—but failed to make the leap into wider, potentially more lucrative consumer markets. As it failed to attract a mass market, district heating became associated with large institutional clients, and the Holly system failed in its efforts to "banish fire completely from our dwellings, office, and factories."[30] This development seems less the result of any technological flaws in the Holly system than of poor public reception and consumer preferences that favored more familiar methods of home heating.

In the end, district heating seemed too radical a change for the American urban household of the late nineteenth century. Home heating would eventually incorporate "networked technologies" of gas and electricity but only after those utility systems and home heating devices designed to take advantage of preexisting source of energy were in place. Even then, as the history of gas and electricity in newer cities like Denver and Kansas City suggest, corporate forces needed extensive marketing campaigns to provide the use of gas and electrical appliances. The changeover to these new household technologies came about as a result of local agents, who could translate the operation of those systems to their neighbors and friends over the years. Only after urban households had incorporated gas or electricity into their lives did they adopt home heating devices drawing on those sources of energy. Decades after technologies like district steam heating tried to work out their kinks in large markets like New York City or Chicago, a multipronged effort was still required to win over consumers and users, and even though district heating is common in other nations, in the American context it is limited largely to institutional markets, like hospital complexes or universities.[31]

The failure of district heating in the industrial household of the late nineteenth century isn't surprising. Historians of household technology across American history argue that in the short-term innovations tend to create familiar systems that resist change. When technological choices are available, families tend to adopt new methods of everyday practice very conservatively. Nineteenth-century consumers might switch their firewood for coal, but they rejected the idea of consuming heat in the form of steam. Instead of purchasing heat as a commodity in and of itself, most urban households preferred to make their own fires. When electrical, gas, and fuel-oil heaters became more popular over the course of the twentieth century, those innovations followed the popularity of existing networked technologies. In fact, coal stoves and furnaces remained in place in many households well into the twentieth century. The urban hearth, which played an important role in America's transition from organic to mineral fuel during the nineteenth-century process of industrialization, no longer drove energy innovations in the twentieth century. The failure of district heating in late nineteenth-century cities suggests that the hearth was no longer a primary location of economic or technological change.[32]

In many ways, the failure of district steam heating brings us full circle to the story of the Franklin stove and the Rumford fireplace. Both of those innovations applied a new form of technology to the hearth, and yet neither revolutionized home heating in the United States during the nineteenth century. District steam heating became, in fact, an idealized system more appropriate to utopian views of the future American city than a real solution to the inadequacies of the coal-burning hearth. In 1894, for example, William T. Stead's *If Christ Came to Chicago* envisioned that city's rebirth in the twentieth century. "A central furnace in each block, fitted with the latest improvements, enabled the municipality to provide heat at a fixed charge for every room in the block," Stead wrote. "In this climate heat is as much a necessity as water, and at City Hall the Heat Department had long been recognized as an indispensable part of the municipal machinery." But these fanciful versions of home heating were not practical in the 1890s, as Americans continued to rely upon the overlapping industrial networks developed over the past half century. Changes in home heating would continue to appear over the next century, but the role of households in driving industrial change was never the same. Instead of leading energy transitions, twentieth-century households followed them: gas, electrical, and oil heating all drew on existing networks of production and distribution and applied them to home heating. Cheap heat, with both its blessings and curses, had finally arrived in the wake of the Industrial Revolution.[33]

Epilogue

In 1873 Philadelphia's Automatic Heating and Lighting Company published a trade catalog listing its wares. Among the many heating devices, the catalog offered a relatively new innovation in home heating, gas heaters. The vast majority of households connected to gas networks used it solely for illumination purposes, as gas lighting dated back to the early nineteenth century in some cities. The conversion of gas to heat, though, was yet another innovation in home heating brought about by the rise of the networked city by the 1870s. Gas heating would become more of a twentieth-century phenomenon, but dealers nonetheless tried to cultivate customers during the late nineteenth century. One of the highlighted devices advertised by the Automatic Heating and Lighting Company was the "Illuminated Chapel," a "pleasing and novel device for Warming and Lighting a sitting room or nursery." The Illuminated Chapel reproduced a small church, with lit windows and spire that could be ordered in a variety of finishes—all iron or a blend of marble and slate to emulate an actual building (fig. E.1). The custom-made heater cost between $100 and $300, depending upon the customer's choice in ornamentation; this price put it well out of the range of most American families. The image accompanying the advertisement nonetheless depicts the ideal in nineteenth-century domestic bliss centered around the hearth: a father reading the newspaper, a mother engrossed in a book, and a young boy playing with a family pet. The only one paying attention to the heater, a young girl, appears to be studying it with some intensity. But even as the family displays various levels of awareness of its presence, the Illuminated Chapel is without question at the center of the picture. It was an example of the kind of hearth that the industrial household could, for the right price, obtain by the turn of the century.[1]

The Illuminated Chapel was by no means a revolutionary device. In fact, its novelty was a deliberate attempt to stand out in what had become a very crowded market for home heating devices. By the outbreak of World War I, for example, American families could choose from a wide range of home heating methods: coal, steam, gas, and of course firewood all served as fuel options for American consumers. Some dealers even pitched petroleum stoves for heating during "those doubtful days—too cold to dispense entirely with fires, too warm to endure a coal fire." The list of devices in which urban households burned that fuel appeared endless. Coal stoves and fireplace grates, hot-air furnaces, radiators that used hot water or steam, and the traditional open fireplace came in a variety of shapes and sizes to fit nearly every situation and every budget. American cities had mitigated the biting fuel crisis that plagued the era of the American Revolution, as the industrial economy made possible all of these methods of home heating. Even as many households retained the idea of a central hearth, the shape, size, and appetite of that hearth had undergone a striking transformation. The famous home economist Ellen Richards illustrates the blend of new and old for the industrial hearth in her turn-of-the-century housekeeping guidebook, *The Cost of Living as Modified by Sanitary Science*. "The heating-plant is the heart of the house for eight or nine months of the year," she observed, "and must be looked after by the most intelligent and responsible person in the house,—one who understands the chemistry of combustion and the mechanics of draft."[2]

Not everyone was pleased with the evolution of the industrial hearth, as the open fireplace, or at least its romantic image, made a comeback in some literary, scientific, and cultural circles of the late nineteenth century. In 1889, for example, poet Joel Benton published a lengthy essay on home heating in *Frank Leslie's Popular Monthly* in which he claimed, "No change has been more revolutionary or pathetic than that which abolished so suddenly the spacious open fireplace." Benton bemoaned the arrival of the stove, the iron furnace, and finally the railroad, which "brought the anthracite coal within everyone's reach, [and] finished the revolution." Stoves, furnaces, and radiators were not, in his mind, to be listed among the benefits of industrial

Figure E.1. The Illuminated Chapel. This gas heater replicated a church within the middle-class home in much the same spirit as early stove designs that contained fancy iron trim or a bust of a famous American. Gas heaters on this scale were relatively rare in the nineteenth century. In this peaceful domestic image, though, only the daughter seems intrigued by the novelty of the device. Automatic Heating and Lighting Company, *Something New: Comforts and Conveniences for Every Family* (Philadelphia: Automatic Heating and Lighting Company, [1873]), 22

society. "Solemnly, as to some symbol of death," he mused, "we go on one by one up to these devices, whose warmth may reach the hands and feet, but which never light the face or warm the heart." Benton also lauded J. Pickering Putnam's *The Open Fireplace in All Ages* (1886), an architectural treatise that advocated the return of fireplaces, as the emphasis upon economy in home heating had all but eliminated them in favor of the "closed stove." Putnam provided the most extensive redesign of the fireplace since Count Rumford, and he hoped to reclaim it from the margins of home heating. Putnam proclaimed that innovations in stove and furnace design over the past century had been "directed to increasing the economy of the heater which is used with the value of economy is best appreciated, to the neglect of the open fireplace," which society had, regrettably, deemed "incurable." Finally, some writers found the industrial hearth's heat simply unsettling. In an 1895 article in *Harper's Bazaar* entitled "Aestheticism in Heat," Clare Brunce denounced the methods by which most middle- and upper-class homes were heated, noting that "either a sickening atmosphere is obtained from steam that pours unremittingly into the rooms, or from some hideous cavern in the wall is sent forth a gaseous heat that is absolutely ruinous to health and to vigor." For many affluent Americans, the open fireplace emerged as a romantic symbol of the cheerfulness and soul of the preindustrial hearth.[3]

The late nineteenth-century industrial economy increased the number of devices available in the home heating market and allowed some Americans the luxury of romanticizing the open fireplace, but the realities of that same economy limited the possibilities of many other consumers. For those living in the crowded tenements, for example, the options were few. Coal still offered economical home heating fuel, but persistent problems in securing it plagued the less affluent residents of American cities. In 1892, one study of the working poor in New York City cited lack of storage space as a major constraint on family budgets. As a result, many tenants purchased coal at the rate of fourteen cents a pail, or fourteen dollars per ton, as opposed to the general retail price of five dollars per ton. New York's new Tenement House Act of 1901 offered an update on tenement construction and maintenance but did little to change home heating. The law provided for "adequate chimneys running through every floor with an open fireplace or grate, or places for a stove," nearly the exact same language used in the original 1867 Tenement Act. Even as elites touted the benefits of wood-burning open fireplaces, poor families resorted to this form of home-heating because they had no other options. An early twentieth-century study of working-class standards of living noted the increase in burning wood—often discarded boxes and construction material—in the homes of the most poor. Although the report offered that "individual economy and extravagance" accounted for some variety in home heating methods, most of New York City's working-class families burned coal if they could afford it. When New York's Tenement Home Department sent representatives to interview residents, the tenants asked for "steam heat and other conveniences." Across the urban landscape of industrial America, the poorest residents still burned wood in their fireplaces, not because it lit their faces or warmed their hearts but out of necessity.[4]

The pervasiveness of the industrial hearth, in all its myriad forms, is evident from the criticism it received from a number of perspectives. Elites might rail against the bland efficiency of a stove or furnace, while less affluent Americans resented their lack of access to the more sophisticated forms of home heating. That the benefits of the industrial hearth should be distributed unevenly across the urban population might come as no surprise. The industrial economy of the late nineteenth century quite notoriously widened the gap between the rich and poor in the United States. In fact, the record of industrialization across the nineteenth century is mixed with regard to equal distribution of its benefits across American society. Nonetheless, staying warm in the winter was generally easier for urban residents by this time. If all but the wealthiest families couldn't afford the Illuminated Chapel, they might at least be able to purchase or rent a cheap coal stove. Paying an outrageous mark-up for fuel was not ideal, but at least it was better than the situation only a few generations earlier, when heating fuel couldn't be had for any price. Like Charles Barnard's quest to go from "hod to mine" in 1874, the present volume has attempted to tell the story of how warmth was brought to the hearth by tracing the development of industrial economy, and like Barnard, is motivated to do so by the simple aim of understanding how these systems actually worked. In our modern times, when heating households usually involves no more than setting the thermostat to a desired temperature, we rarely probe the matter as thoroughly as Barnard did in 1874 or reflect much at all on the way we heat our homes.

So what does appreciating the complexity of changes in home heating during the nineteenth century tell us? At a very broad level, it forces us to examine the connections that we all have to wider networks of production, distribution, and consumption. One of the lasting legacies of the industrial transformation of the American economy is that it gave rise to these complex systems that, on the one hand, allowed an unprecedented increase in the material standards of living but, on the other, made the residents of urban areas more and more dependent upon those networks for daily comforts. The industrial hearth thus became a vital connection between households and wider economic forces, and it offers several tangible examples of widespread transformations in society that occurred in response to the simple task of staying warm in the winter. This challenge, for example, created the first durable consumer good in American manufacturing with the rise of the stove. It helped spur the nation's coal trade, upon which the American industrial economy grew to become the most powerful in the world and allowed cities like New York, Philadelphia, and Chicago to develop into manufacturing powerhouses. Within these cities, however, residents negotiated the rapid changes in home heating according to their economic standing, and new markets in mineral fuel often heightened the difficulties of the urban poor to secure it. Whereas wealthy consumers could afford the luxury of purchasing a large amount of coal in the summer months and laying up a season's supply, the less affluent paid more for heating fuel when they bought it in smaller increments. Both the wealthy and poor complained about the ubiquitous and necessary agents of home heating in industrial America, the coal dealers. Home heating created the first cartel in big business and in Franklin Gowen produced an example of a very powerful entrepreneur who beat the Molly Maguires but couldn't beat low prices. Finally, home heating tested the upper limits of the industrial reorganization of American cities. Urban households might be willing to tinker with how they kept warm over the long winter nights, but they didn't want to surrender control over combustion or abandon all familiar vestiges of the traditional hearth. The industrial economy's impact on home heating offers us a way to understand how people negotiated these revolutionary changes in everyday life over a few short generations.

Another legacy of this era is the American dependency on fossil fuels. Keeping the flow of coal into the industrial households of the nineteenth century anticipated the need to keep oil flowing into the automobile-driven economy of the twentieth century. During the oil crisis of 1973, when petroleum-exporting nations of the Middle East placed an embargo on imports to the United States, the shock to the American economy was immense. Prices skyrocketed, fuel shortages ensued, and gasoline was rationed. The long-term impact on global political economy was equally profound, as Middle Eastern nations found oil an effective means of leveling the playing field in international relations. This crisis served as wake-up call for the United States. President Richard Nixon appeared on television on November 7, 1973, to explain to Americans that "our energy demands have begun to exceed available supplies" because "our economy has grown enormously and what were once considered luxuries are now considered necessities." Among those comforts of life were fuel-oil furnaces, a twentieth-century innovation designed to take advantage of what seemed to be the fuel of the future, petroleum. Of course, the scarcity of gasoline for automobiles was President Nixon's major concern during the energy crisis of 1973, but he also targeted home heating by asking everyone to turn down thermostats by at least six degrees. In language that P. T. Barnum's antistove churchgoers surely would have appreciated, Nixon championed a brisk indoor atmosphere. "Incidentally, my doctor tells me that in a temperature of 66 to 68 degrees," he announced, "you are really more healthy than when it is 75 to 78, if that is any comfort."[5]

The crisis of the 1970s highlighted the dependence of the American economy on foreign imports of oil, but it was not the first time that a threat to cheap energy and home heating spurred a major shift in policy. Beginning in 1900, the decade-old United Mine Workers of America organized more than one hundred thousand miners in Pennsylvania's anthracite fields into a series of strikes, cities faced a devastating shortage of coal. The largest of these, the Great Anthracite Coal Strike of 1902, threatened to deprive American cities of an everyday commodity. That strike lingered all summer, and by September of that year, hard coal prices in large cities had quadrupled to twenty dollars a ton. Talk of a "coal famine" in the upcoming winter months dominated the headlines. "It is no use now to say whose fault it is," one frustrated reader from New Jersey wrote the *New York Times*. "We must decide on the best course to mitigate as much as possible the suffering, sickness, disorder, and loss which in any event must bear heavily upon us for some months." Some pundits called for President Theodore Roosevelt to nationalize the anthracite trade; others demanded that the US military occupy the region and force miners to work. In the end, President Roosevelt declined to pursue either of these radical options but nonetheless sent messages to both labor and business leaders demanding a meeting in Washington "in regard to the failure of the coal supply which has become a matter of vital concern to the whole nation." The creation of the Anthracite Strike Coal Commission, an ad hoc committee made up of government, labor, and business representatives, ended the coal famine and in the following year issued a report that at least temporarily ended the labor strife in the anthracite regions. Most historians summed up the achievement of Roosevelt's commission as offering a "square deal" for workers; of perhaps greater importance was the fact that the specter of a widespread "coal famine" pushed the nation's leaders into uncharted political territory and changed the role of government. Fossil fuel dependence and the assumption that cheap energy will always be available, first forged in the industrial hearths of the nineteenth century, had metastasized in the American way of life by the time of the Great Anthracite Coal Strike of 1902. In many ways, it is still with us today.[6]

The rise of coal in home heating also reveals a number of important themes that inform both past and future decisions about energy use. First and foremost, it stresses the importance of disaggregating the experience of poor and wealthy consumers in energy transitions. The rise of the industrial hearth revealed that the approach to heating fuel use differed widely by class status. This is not to say that all wealthy Americans made the transition instantly, while those in the lower income brackets were forced kicking and screaming into using coal or steam to heat their households. Nor does it suggest that poorer consumers could not adopt cutting-edge technology or embrace new commodities. It does, however, point toward the sustained effort required to change energy consumption and the different levels at which this process must occur. Anthracite reached the affluent through testimonials, word of mouth, and trial runs; it found its way to the poor through philanthropic subsidies. The complexities involved in this energy transition should give us a working model with which to understand contemporary energy regimes. The various ways in which anthracite made its way into urban hearths suggest that in the current debate over alternatives to fossil fuel use there is no magic bullet that will provide a rapid transition to renewable resources. Of course, the circumstances and technological challenges are different, but the challenges of convincing a wide variety of consumers to abandon one relatively stable practice for a new and unfamiliar one has many precedents. If a "green" revolution in renewable energy use is to occur in modern times, we would do well to understand the fossil fuel–based "black" one that occurred in American cities during the nineteenth century. To do so, we must look beyond the relatively easy process of coming up with the modern-day equivalent of the Franklin stove or Rumford fireplace—good ideas that had only a limited impact on economic change—and instead think about the wider systems necessary to tackle the problem of energy production and consumption in the future.

NOTES

Abbreviations

CHS	Chicago Historical Society, Chicago, IL
HML	Hagley Museum and Library, Greenville, DE
HSP	Historical Society of Pennsylvania, Philadelphia, PA
MHS	Massachusetts Historical Society, Boston, MA
NYHS	New-York Historical Society, New York, NY
UVA	University of Virginia, Charlottesville, VA

Prologue

1. Mary Caroline Crawford, *Old Boston Days and Ways: From the Dawn of the Revolution until the Town Became a City* (Boston: Little, Brown, 1909), 172–174; William Smith diary entry of 5 Jan. 1780, quoted in I. N. Phelps Stokes, *The Iconography of Manhattan Island, 1498–1909*, vol. 5 (New York: Robert Dodd, 1926), 1100; David Ludlum, *Early American Winters, 1604–1820* (Boston: American Meteorological Society, 1966), 111–117.

2. Billy Smith, *The "Lower Sort": Philadelphia's Laboring People, 1750–1800* (Ithaca, NY: Cornell University Press, 1990), 101–106; Susan Klepp, "The Working Poor in Philadelphia: Gender and Infant Mortality," in *Down and Out in Early America*, ed. Billy G. Smith (University Park: Pennsylvania State University Press, 2004), 73; Susan Klepp, ed., *"The Swift Progress of Population": A Documentary and Bibliographic Study of Philadelphia's Growth, 1642–1859* (Philadelphia: American Philosophical Society, 1991), passim; Harvard officials quoted in William B. Meyer, "Harvard and the Heating Revolution," *New England Quarterly* 77 (2004): 591.

3. Lawrence Wright, *Home Fires Burning: The History of Domestic Heating and Cooking* (London: Routledge, 1964), 9–82; John E. Crowley, *The Invention of Comfort: Sensibilities and Design in Early Modern Britain and Early America* (Baltimore: Johns Hopkins University Press, 2001), 3–36. The term *household* (as opposed to *family*) is used as the unit of analysis in this book is not individuals related by marriage or blood but people who cohabitate. That is, people living under the same roof constitute a "household," even though they might not consider themselves "family." This is a necessary, if blunt, shorthand; the idea of what exactly constituted a "household" and who made the main decisions for it was in flux throughout this time period. See

Carole Shammas, *A History of Household Government in America* (Charlottesville: University of Virginia Press, 2002), 1–23.

4. Cadwallader Colden, "Account of the Climate and Diseases of New-York," *American Medical and Philosophical Register* 1 (Jan. 1811): 309; William B. Meyer, *Americans and Their Weather* (New York: Oxford University Press, 2000), 32–33; Seth Rockman, *Scraping By: Wage Labor, Slavery, and Survival in Early Baltimore* (Baltimore: Johns Hopkins University Press, 2009), 179–180.

5. Benjamin Franklin, *An Account of the Newly Invented Pennsylvanian Fire-Place: Wherein Their Construction and Manner of Operation Is Particularly Explained* (Philadelphia, 1744), 1, 18, 23, 25.

6. Ibid., 6; Joyce Chaplain, *The First Scientific American: Benjamin Franklin and the Pursuit of Genius* (New York: Basic Books, 2006), 84–92.

7. Graf von Rumford [Benjamin Thompson], "Of the Fundamental Principles on Which General Establishments for the Relief of the Poor may be formed in all Countries," in *Essays, Political, Economical, and Philosophical*, vol. 1 (Boston: Manning and Loring, 1798), 178; Wright, *Home Fires Burning*, 113–118; Sanborn C. Brown, *Benjamin Thompson, Count Rumford* (Cambridge, MA: MIT Press, 1979), 163–176.

8. Graf von Rumford [Benjamin Thompson], "Of Chimney Fire-Places, with Proposals for Improving them to Save Fuel; to Render Dwelling Houses More Comfortable and Salubrious, and Effectually to Prevent Chimnies from Smoking," in *Essays, Political, Economical, and Philosophical*, vol. 1, 305, 320–321, 364–365; "Review of Count Rumford's Second Essay," *Weekly Magazine* 2 (12 May 1798): 37.

9. See, for example, Samuel J. Edgerton, "Supplement: The Franklin Stove," in I. Bernard Cohen, *Benjamin Franklin's Science* (Cambridge, MA: Harvard University Press, 1990), 199–211; Chaplin, *The First Scientific American*, 92; Walter Isaacson, *Benjamin Franklin: An American Life* (New York: Simon and Schuster, 2003), 132; Edmund S. Morgan, *Benjamin Franklin* (New Haven, CT: Yale University Press, 2002), 11.

10. Joel Mokyr, *The Enlightened Economy: An Economic History of Britain, 1700–1850* (New Haven, CT: Yale University Press, 2009), 5. The literature on the definition of "technology" is vast. For a brief overview of the historical definition and its relationship to economic change, see Leo Marx, "Technology: The Emergence of a Hazardous Concept," *Technology and Culture* 51 (2010): 561–577.

11. There are a multitude of works that deal with industrialization in the United States. For a good, broad overview, see Walter Licht, *Industrializing America: The Nineteenth Century* (Baltimore: Johns Hopkins University Press, 1995). For more on the transition from an organic to a mineral fuel economy in the United States, see Thomas Andrews, *Killing for Coal: America's Deadliest Labor War* (Cambridge, MA: Harvard University Press, 2008), 35–86; Christopher Jones, "A Landscape of Energy Abundance: Anthracite Coal Canals and the Roots of American Fossil Fuel Dependence, 1820–1860," *Environmental History* 15 (2010): 449–484; E. A. Wrigley, *Energy and the English Industrial Revolution* (New York: Cambridge University Press, 2010).

CHAPTER ONE: How the Industrial Economy Made the Stove

1. David Hackett Fischer, *Liberty and Freedom: A Visual History of America's Founding Ideas* (New York: Oxford University Press, 2004), 27; Arthur Schlesinger, "Liberty Tree: A Genealogy," *New England Quarterly* 25 (1952): 434–458.

2. *New England Chronicle*, 31 Aug. 1775; *Massachusetts Gazette and Boston Weekly News-Letter*, 22 Feb. 1776; Fischer, *Liberty and Freedom*, 29.

3. For more on the notion of the stove as the first major consumer durable, see Howell J. Harris, "Inventing the US Stove Industry, c.1815–1875: Making and Selling the First Universal Consumer Durable," *Business History Review* 82 (2008): 701–733.

4. R. V. Reynolds and Albert H. Pierson, *Fuel Wood Used in the United States, 1630–1930*, Circular no. 641 (Washington, DC: United States Department of Agriculture, 1942), 1–4; Arthur Cole, "The Mystery of Fuel Wood Marketing in the United States," *Business History Review* 44 (1970): 343–344; Michael Williams, *Americans and Their Forests: A Historical Geography* (New York: Cambridge University Press, 1989), 133–134.

5. Carl Bridenbaugh, *Cities in the Wilderness: The First Century of Urban Life in America, 1625–1742* (New York: Ronald, 1938), 11–12, 152; *Minutes of the Common Council of the City of New York, 1675–1776*, vol. 1, *1675–1696* (New York: Dodd, Mead, 1905), 146.

6. Graham Russell Hodges, *New York City Cartmen, 1667–1850* (New York: New York University Press, 1986), 51–52; *Minutes of the Common Council of the City of New York, 1675–1776*, vol. 7, *1755–1765* (New York: Dodd, Mead, 1905), 46, 148–149; Carl Bridenbaugh, *Cities in Revolt: Urban Life in America, 1743–1776* (New York: Alfred A. Knopf, 1955), 27, 233–234.

7. Williams, *Americans and Their Forests*, 136; John Cushing, "To Destroy Canker-Worms, and Prevent the Blazing of Grain," *American Museum* 9 (1791): 45; *Weekly Magazine* (Philadelphia), 10 Mar., 19 May 1798.

8. *Minutes of the Common Council of the City of New York, 1784–1831*, vol. 3, *1801–1805* (New York: M. B. Brown, 1917), 148–150, 608–609; A Law to Regulate the Sale of Fire-Wood and to Regulate Carts, Cartmen, &c. &c. in the City of New York (passed 1814), 3–15; *Poulson's American Daily Advertiser* (Philadelphia), 28 Jan. 1801; James Mease, *The Picture of Philadelphia* (Philadelphia: B. and T. Kite, 1811), 126–128; Hodges, *New York City Cartmen*, 133.

9. Meyer, *Americans and Their Weather*, 72–73; Elizabeth Blackmar, *Manhattan for Rent, 1785–1850* (Ithaca, NY: Cornell University Press, 1989), 47–48, 70–71; Donna J. Rilling, *Making Houses, Crafting Capitalism: Builders in Philadelphia, 1790–1850* (Philadelphia: University of Pennsylvania Press, 2001), 10, 47–50.

10. Benjamin Franklin, "Description of a New Stove for Burning of Pitcoal, and Consuming all its Smoke," *Transactions of the American Philosophical Society Held at Philadelphia, for Promoting Useful Knowledge*, vol. 2 (Philadelphia: Robert Aitken, 1786), 71; Samuel J. Edgerton, "Heating Stoves in Eighteenth-Century Philadelphia," *Bulletin of the Association for Preservation Technology* 3 (1971): 25–26; Brooke Hindle, *David Rittenhouse* (Princeton, NJ: Princeton University Press, 1964), 247; John C.

Wills, "The Politics of Taste in the New Republic: The Decorative Elaboration of the Philadelphia Household, 1780–1820," (Ph.D. diss., University of Michigan, 1994), 206–259.

11. "Advertisement," *Transactions of the American Philosophical Society, Held at Philadelphia, for Promoting Useful Knowledge*, 4 (1799): v; "On Fireplaces" 16 Oct. 1796, box 5, Manuscript Communications, American Philosophical Society Archives, Philadelphia; Sidney Hart, "'To Encrease the Comforts of Life': Charles Willson Peale and the Mechanical Arts," *Pennsylvania Magazine of History and Biography* 110 (1986): 336–338; *Poulson's American Daily Advertiser* (Philadelphia), 1 Feb. 1803.

12. H. H. Manchester, *The Evolution of Cooking and Heating* (Troy, NY: Fuller and Warren, 1917), 10–12; Robert Bruegmann, "Central Heating and Forced Ventilation: Origins and Effects on Architectural Design," *Journal of the Society of Architectural Historians* 37 (1978): 144, 146–147; Thomas Tredgold, *Principles of Warming and Ventilating Public Buildings, Dwelling-Houses, Manufactories, Hospitals, Hot-Houses, Conservatories, &c.*, 2nd ed. (London: Josiah Taylor, 1824), 19; Charles Hood, *A Practical Treatise on Warming Buildings by Hot Water; on Ventilation, and the Various Methods of Distributing Artificial Heat* (London: Whittaker, 1844), 230.

13. Robert Bruegmann, "Central Heating and Forced Ventilation: Origins and Effects on Architectural Design," *Journal of the Society of Architectural Historians* 37 (1978): 144–150; Daniel Pettibone, *Description of the Improvements of the Rarifying Air-Stove, for Warming and Ventilating Hospitals, Churches, Colleges, Courts of Justice, Dwellinghouses, Hot or Greenhouses, Manufactories, Banks, Barracks, Ships, &c. &c.* (Philadelphia, 1810), 19–25; Lillian B. Miller, ed., *The Selected Papers of Charles Willson Peale and His Family*, vol. 2, part 1, *Charles Willson Peale: The Artist as Museum Keeper, 1791–1810* (New Haven, CT: Yale University Press, 1988), 210; Benjamin L. Walbert, "The Infancy of Central Heating in the United States, 1803–1845," *Bulletin of the Association for Preservation Technology* 3 (1971): 78. Howell Harris makes a persuasive argument that nonhousehold settings—those areas most likely to have central heating systems—also spurred the use of stoves during the Early Republic. See Howell J. Harris, "Conquering Winter: US Consumers and the Cast-Iron Stove," *Building Research & Information* 36 (2008): 343.

14. William B. Meyer, "Harvard and the Heating Revolution," *New England Quarterly* 77 (2004): 590; Billy G. Smith, *The "Lower Sort": Philadelphia's Laboring People, 1750–1800* (Ithaca, NY: Cornell University Press, 1990), 101–106; Ellis Paxson Oberholtzer, *Philadelphia: A History of the City and Its People*, vol. 1 (Philadelphia: S. J. Clarke, 1912), 400; *Poulson's American Daily Advertiser* (Philadelphia), 25, 28 Jan. 1805; Eliza Cope Harrison, ed., *Philadelphia Merchant: The Diary of Thomas P. Cope, 1800–1851* (South Bend, IN: Gateway Editions, 1978), 178; Charles Peirce, *A Meteorological Account of the Weather in Philadelphia, from January 1, 1790 to January 1, 1847, Including Fifty-Seven Years* (Philadelphia: Lindsay and Blakiston, 1847), 37–38; Mease, *The Picture of Philadelphia*, 40–41; Dorothy C. Barck, ed., *Letters from John Pintard to His Daughter, Eliza Noel Pintard Davidson, 1816–1833*, vol. 1, *1816–1820* (New York: New-York Historical Society, 1940), 56.

15. Josephine H. Peirce, *Fire on the Hearth: The Evolution and Romance of the Heating Stove* (Springfield, MA: Pond-Ekberg, 1951), 93; Edgerton, "Heating Stoves in Eighteenth-Century Philadelphia," 17–18, 26–28.

16. Peirce, *Fire on the Hearth*, 38–39; James Moore Swank, *The History of Iron Manufacture in All Ages; and Particularly the United States from the Colonial Times to 1891* (Philadelphia: American Iron and Steel Association, 1892), 179; George L. Heiges, *Henry William Stiegel: The Life Story of a Famous American Glass-Maker* (Manheim, PA: published by the author, 1937), 4–8, 16–18; Frederick William Hunter, *Stiegel Glass* (New York: Dover, 1950), 5–30, 60, 94.

17. Priscilla Brewer, *From Fireplace to Cookstove: Technology and the Domestic Ideal in America* (Syracuse, NY: Syracuse University Press, 2000), 37; Peirce, *Fire on the Hearth*, 39–48; Edwin A. Barber, "Cast Iron Stoves of the Pennsylvania Germans," *Bulletin of the Pennsylvania Museum* 13 (1915): 19–23. A wide range of religious scenes were depicted on early American stove plates. For a comprehensive catalog of them, including photographs, see Henry C. Mercer, *The Bible in Iron; or, Pictured Stoves and Stove Plates of the Pennsylvania Germans* (Doylestown, PA: Bucks County Historical Society, 1941).

18. Tredgold, *Principles of Warming and Ventilating*, 4–5, 8; John E. Crowley, *The Invention of Comfort: Sensibilities and Design in Early Modern Britain and Early America* (Baltimore: Johns Hopkins University Press, 2001), 190; Brewer, *From Fireplace to Cookstove*, 22–25, 36. Gabrielle Lanier argues that the adoption of stoves to Georgian architectural traditions represents a "creolization" of the use of stoves that might have been obvious to visitors inside a house but less apparent from the exterior of the home visible to the public. Gabrielle M. Lanier, *The Delaware Valley in the Early Republic: Architecture, Landscape, and Regional Identity* (Baltimore: Johns Hopkins University Press, 2005), 64–68.

19. *United States Gazette* (Philadelphia), 28 Nov. 1814; Temin, *Iron and Steel in Nineteenth-Century America: An Economic Inquiry* (Cambridge, MA: MIT Press, 1964), 37–38, 83–84; Walker, *Hopewell Village: A Social and Economic History of an Iron-Making Community* (Philadelphia: University of Pennsylvania Press, 1966), 141; Williams, *Americans and Their Forests*, 107–109; Arthur D. Pierce, *Iron in the Pines: The Story of New Jersey's Ghost Towns and Bog Iron* (New Brunswick, NJ: Rutgers University Press, 1957), 103–104.

20. George Rogers Taylor, *The Transportation Revolution, 1815–1860* (New York: Holt, Rinehart, and Winston, 1951), 132–138; Carol Sheriff, *The Artificial River: The Erie Canal and the Paradox of Progress* (New York: Hill and Wang, 1996), 52–78; John Lauritz Larson, *Internal Improvement: National Public Works and the Promise of Popular Government in the Early United States* (Chapel Hill: University of North Carolina, 2001), 71–107.

21. Harold C. Livesay, "Marketing Patterns in the Antebellum Iron Industry," *Business History Review* 45 (1971): 274–275; Glenn Porter and Harold C. Livesay, *Merchants and Manufacturers: Studies in the Changing Structure of Nineteenth-Century Marketing* (Baltimore: Johns Hopkins Press, 1971), 53–54; Howell J. Harris, "Inventing

the US Stove Industry," 706; Robert Brookhouse to Samuel Wright, 18 Mar. 1822, John Bradbury & Co. to Samuel Wright, 24 Dec. 1831, Wright Family Papers, HML.

22. Edgerton, "Heating Stoves in Eighteenth-Century Philadelphia," 22; Brewer, *From Fireplace to Cookstove*, 63–64; Harris, "Inventing the US Stove Industry," 710–711; B. Zorina Khan, *The Democratization of Invention: Patents and Copyrights in American Economic Development, 1790–1920* (New York: Cambridge University Press, 2005), 106–127, quote from p. 106; David R. Meyer, *The Roots of American Industrialization* (Baltimore: Johns Hopkins University Press, 2003), 66–69.

23. Affidavits dated 25 Jan. 1816 and 7 Feb. 1834, James Wilson to Morse & Son, 6 Apr. 1839, and Thomas Jones to James Wilson, 25 July 1842, 25 Aug. 1842, all in James Wilson Collection, NYHS.

24. Codman Hislop, *Eliphalet Nott* (Middletown, CT: Wesleyan University Press, 1971), 257–269; James Wilson to Nott & Co., 3 Sept. 1833, James Wilson Papers, NYHS; Howell Harris, "'The Stove Trade Needs Change Continually': Designing the First Mass-Market Consumer Durable, ca. 1810–1930," *Winterthur Portfolio* 43 (2009): 365–406; Tammis Kane Groft, *Cast with Style: Nineteenth-Century Cast-Iron Stoves from the Albany Area* (Albany, NY: Albany Institute of History and Art, 1984), 90.

25. Brewer, *From Fireplace to Cookstove*, 77–80; *Longworth's American Almanac, 1815* (New York: Directory Office, 1815); *Longworth's American Almanac, 1830* (New York: Directory Office, 1830); Jordan L. Mott, *Description and Design of Mott's Patented Articles, Secured by 27 Patents* (New York: Daniel Adee, 1841).

26. Harris, "A Fable of Progress," 14–15; *Doggett's New-York City Directory for 1845 and 1846* (New York: John Doggett Jr., 1846).

27. Groft, *Cast With Style*, 16–17, 23–24; J. Leander Bishop, *A History of American Manufactures from 1608 to 1860*, vol. 1 (Philadelphia: Edward Young, 1864), 625; Victor Clark, *History of Manufactures in the United States*, vol. 1, *1607–1860* (New York: McGraw-Hill, 1929), 503.

28. Clark, *History of Manufactures*, 503–504; Groft, *Cast with Style*, 16; Temin, *Iron and Steel in Nineteenth-Century America*, 38–39; Harris, "Conquering Winter," 347; Harris, "Inventing the US Stove Industry," 723–728.

29. Phineas Taylor Barnum, *The Autobiography of P. T. Barnum* (London: Ward and Lock, 1855), 20; Robert Cray, "Heating the Meeting: Pro-Stove and Anti-Stove Dynamic in Church Polity, 1783–1830," *Mid-America: An Historical Review* 76 (1994): 93–107.

30. Dorothy S. Brady, "Relative Prices in the Nineteenth Century," *Journal of Economic History* 24 (1964): 180; Brewer, *From Fireplace to Cookstove*, 78–80.

CHAPTER TWO: How Mineral Heat Came to American Cities

1. Eliza Leslie, *The House Book; or, A Manual of Domestic Economy* (Philadelphia: Carey and Hart, 1840), 129, 136–138.

2. On the early rise of coal in Pennsylvania, see Clifton K. Yearley, *Enterprise and Anthracite: Economics and Democracy in Schuylkill County, 1820–1875* (Baltimore: Johns Hopkins Press, 1961); Alfred D. Chandler, Jr., "Anthracite Coal and the Begin-

nings of the Industrial Revolution in the United States," *Business History Review* 46 (1972): 141–181; and H. Benjamin Powell, *Philadelphia's First Fuel Crisis: Jacob Cist and the Developing Market for Pennsylvania Anthracite* (University Park: Pennsylvania State University Press, 1978). For its wider adoption, consult Sam H. Schurr and Bruce C. Netschert, *Energy in the American Economy, 1850–1975: An Economic Study of Its History and Prospects* (Baltimore: Johns Hopkins Press, 1960); Frederick Binder, *Coal Age Empire: Pennsylvania Coal and Its Utilization to 1860* (Harrisburg: Pennsylvania Historical and Museum Commission, 1974); Barbara Freese, *Coal: A Human History* (Cambridge, MA: Perseus, 2003); and Alfred Crosby, *Children of the Sun: A History of Humanity's Unappeasable Appetite for Energy* (New York: W. W. Norton, 2006).

3. The significance of the domestic side of American energy transitions is a major theme in Christopher Jones, "The Carbon-Consuming Home: Residential Markets and Energy Transitions," *Enterprise & Society* 12 (Dec. 2011): 790–823. See also Martin Melosi, "Energy Transitions in the Nineteenth-Century Economy," in *Energy and Transport. Historical Perspectives on Policy Issues*, ed. George H. Daniels and Mark H. Rose (Beverley Hills, CA: Sage, 1982), 55–69; Vaclav Smil, *Energy in World History* (Boulder, CO: Westview, 1994); David Nye, *Consuming Power: A Social History of American Energies* (Cambridge, MA: MIT Press, 1998). Two very long-term perspectives are in Rolf Peter Sieferle, *The Subterranean Forest: Energy Systems and the Industrial Revolution* (London: White Horse, 2010); and Roger Fouquet, *Heat, Power and Light: Revolutions in Energy Services* (Northampton, MA: Edward Elgar, 2010).

4. Howard Eavenson, *The First Century and a Quarter of American Coal Industry* (Pittsburgh: privately printed, 1942), 442; Harry Heth to David Meade Randolph, 22 June 1814, Henry Heth Papers, UVA.

5. "Duty on Coal: Communicated to the House of Representatives, February 9, 1798," and "Duty on Coal: Communicated to the House of Representatives, February 4, 1802," *American State Papers. Documents, Legislative and Executive of the Congress of the United States* 5 (Washington: Gales and Seaton, 1832), 553, 729.

6. F. W. Taussig, *The Tariff History of the United States*, 8th ed. (New York: G. P. Putnam's Sons, 1931), 17; Arthur Cole, *Wholesale Commodity Prices in the United States, 1700–1861* (Cambridge, MA: Harvard University Press, 1938), 164, 178; Curtis P. Nettels, *The Emergence of a National Economy, 1775–1815* (New York: Holt, Rinehart, and Winston, 1962), 324–335.

7. Harry Heth to Thomas Railey & Brother, 16 June 1817, Henry Heth Papers, UVA. On the struggles to exploit eastern Virginia's coalfields, see Sean Patrick Adams, "Pits of Frustration: The Failed Transplant of British Mining Methods in Antebellum Virginia," in *Technology, Innovation, and Southern Industrialization: From the Antebellum Era to the Computer Age*, ed. Susanna Delfino and Michele Gillespie (Columbia: University of Missouri Press, 2008), 41–67.

8. *An Enquiry into the Chymical Character and Properties of that Species of Coal, Lately Discovered at Rhode Island* (Boston: Snelling and Simon, 1808), 4, 9; RICC, *Observations on the Rhode Island Coal, and Certificates with Regard to Its Qualities, Value, and Various Uses* (Boston: n.p., 1814), 7, 15–16.

9. Eavenson, *American Coal Industry*, 502; Gardner Dagget to Samuel Waldron, 29 May 1819, Thomas Cary to Samuel Waldron, 7 Dec. 1823, H. D. Sedgwick to Samuel Waldron, 3 May 1825, all in Samuel Waldron Papers, MHS.

10. RICC, *Observations on the Rhode Island Coal*, 8; *An Address to the Inhabitants of Rhode-Island, on the Subject of their Coal Mines* (New York: J. Seymour, 1825), 9; H. D. Sedgwick, *Circular* (Boston: W. L. Lewis, 1827), 6, 14; John Rynex to Samuel Waldron, 25 June 1835, Samuel Waldron Papers, MHS.

11. Christopher Jones, "Energy Landscapes: Coal Canals, Oil Pipelines, and Electricity Transmission Wires in the Mid-Atlantic, 1820–1930," (PhD diss., University of Pennsylvania, 2009), 91–157.

12. Evans knew about the timing of fuel markets firsthand, as he aggressively advertised and sold 600 cords of oak and 120 cords of pine in the severe winter of 1809. See *Relfs Philadelphia Gazette, and Daily Advertiser*, 14 Feb. 1809; *Poulson's American Daily Advertiser*, 16 Feb. 1809; *Philadelphia Political and Commercial Register*, 20 Feb. 1809; *Address of the President and Managers of the Schuylkill Navigation Company, to the Stockholders and to the Publick in General* (Philadelphia: United States Gazette, 1817), 7; Andrew Schocket, *Founding Corporate Power in Early National Philadelphia* (DeKalb: Northern Illinois University Press, 2007), 177–181.

13. *Facts Illustrative of the Character of the Anthracite, or Lehigh Coal, Found in the Great Mines at Mauch Chunk, in Possession of the Lehigh Coal and Navigation Company, with Certificates from Various Manufacturers, and Others, Proving Its Decided Superiority over Every Other Kind of Fuel* (Philadelphia: Solomon W. Conrad, 1827), 16–17.

14. Frederick M. Binder, "Anthracite Enters the American Home," *Pennsylvania Magazine of History and Biography* 82 (Jan. 1958): 92; "History of the Introduction of Anthracite Coal into Philadelphia. By Erskine Hazard, Esq., Communicated to the Society, Feb. 5, 1827; and a Letter from Jesse Fell, Esq. of Wilkesbarre, on the Discovery and First Use of Anthracite in the Valley of Wyoming," in Historical Society of Pennsylvania, *Memoirs of the Historical Society of Pennsylvania*, 14 vols. (Philadelphia: McCarty and Davis, 1826–1895), 2:163; Lehigh Coal and Navigation Company, "Extract of a Letter from Philadelphia, dated October 13, 1825," in *Facts Illustrative of the Character of the Anthracite*, 16–17; John E. Crowley, *The Invention of Comfort: Sensibilities and Design in Early Modern Britain and Early America* (Baltimore: Johns Hopkins University Press, 2001), 292; Nicholas B. Wainwright, ed., *A Philadelphia Perspective: The Diary of Sidney George Fisher Covering the Years 1834–1871* (Philadelphia: Historical Society of Pennsylvania, 1967), 314.

15. James Mease and Thomas Porter, *Picture of Philadelphia, Giving an Account of Its Origin, Increase and Improvements in Arts, Sciences, Manufactures, Commerce and Revenue with a Compendious View of Its Societies, Literary, Benevolent, Patriotic, and Religious*, 2 vols. (Philadelphia: Robert Desilver, 1831), 2:58–59; *Niles' Register*, 20 June 1835.

16. T. H. Breen, *The Marketplace of Revolution: How Consumer Politics Shaped American Independence* (New York: Oxford University Press, 2004); Joyce Appleby, "Consumption in Early Modern Social Thought," in *Consumption and the World of*

Goods, ed. John Brewer and Roy Porter (New York: Routledge, 1993), 162–173; Crowley, *The Invention of Comfort*, 171–190. For an overview of the idea of "consumption" in history, see Susan Strasser, "Making Consumption Conspicuous: Transgressive Topics Go Mainstream," *Technology and Culture* 43 (2002): 755–770.

17. Sidney W. Mintz, *Sweetness and Power: The Place of Sugar in Modern History* (New York: Viking, 1985). For a discussion of the "democratization" of sweets in the American context, see Wendy A. Woloson, *Refined Tastes: Sugar, Confectionary, and Consumers in Nineteenth-Century America* (Baltimore: Johns Hopkins University Press, 2002), 2–16. Paul G. E. Clemens downplays the notion of a consumer "revolution" and highlights the regional character of that trend by tracing the very gradual change in consumer goods in Mid-Atlantic households. See Paul G. E. Clemens, "The Consumer Culture of the Middle Atlantic, 1760–1820," *William and Mary Quarterly*, 3rd ser., 62 (Oct. 2005): 577–624.

18. Jeanne Boydston, *Home and Work: Housework, Wages, and the Ideology of Labor in the Early Republic* (New York: Oxford University Press, 1990), 105; Supply Clap Thwing to Susan Ela, 13 Sept. 1849, Supply Clap Thwing Letterbooks, MHS. The impact of users on technology is outlined in the various essays of Nelly Oudshoorn and Trevor Pinch, eds., *How Users Matter: The Co-Construction of Users and Technology* (Cambridge, MA: MIT Press, 2003).

19. Robert Roberts, *The House Servant's Directory; or, A Monitor for Private Families* (Boston, Munroe and Francis, 1827), 159; Amateur, *Directions for the Use of Anthracite Coal* ([New York, 1835]), 7; Leslie, *The House Book*, 135; Denison Olmsted, *Observations on the Use of Anthracite Coal* (Cambridge, MA, 1836), 4.

20. Amateur, *Directions for the Use of Anthracite Coal*, 5, 8, 12; *Niles' Register*, 1 Nov. 1834.

21. The Fuel Savings Society estimated that a poor family in Philadelphia consumed about 2.5 cords of wood in the winter season, which cost around fifteen dollars in 1831. The equivalent heating value of anthracite—they calculated about two tons—would cost around nine dollars in 1831. But it is important to remember that a conversion from wood to anthracite fuel would require an initial investment in either an anthracite grate or stove, which the society did not included in its report. See Fuel Savings Society of the City and Liberties of Philadelphia, *A History of the Fuel Savings Society of the City and Liberties of Philadelphia from its Organization to 1871* (Philadelphia, 1875), 7–8.

22. UBA, *First Annual Report of the Executive Board of the Union Benevolent Association* (Philadelphia, 1832), 3; minutes of the executive board, 6 Sept. 1832 and 28 Oct. 1833, in "Minutes of the Executive Board, 1831–1854," UBA Records, HSP; W. H. Keating, *Considerations upon the Art of Mining. To Which Are Added, Reflections on Its Actual State in Europe, and the Advantages Which Would Result from an Introduction of This Art into the United States* (Philadelphia, 1821).

23. Binder, "Anthracite Enters the American Home," 94; Cornelius Stevenson Receipt Book, HSP.

24. *Relfs Philadelphia Gazette, and Daily Advertiser*, 19 Jan. 1831; *Niles' Register*, 16

July 1831; Fuel Savings Society, *A History of the Fuel Savings Society*, 9; UBA, *Union Benevolent Association, 1831–1881: Fifty Years of Work among the Poor of Philadelphia; Historical Sketch of the First Half-Century of the Union Benevolent Association* (Philadelphia, 1881), 25; Priscilla Clement, "Nineteenth-Century Welfare Policy, Programs, and Poor Women: Philadelphia as a Case Study," *Feminist Studies* 18 (1992): 38. Estimates on the cost of stoves come from Priscilla Brewer, who notes the "sticker shock" that most antebellum consumers faced when purchasing a stove. Brewer, *From Fireplace to Cookstove: Technology and the Domestic Ideal in America* (Syracuse, NY: Syracuse University Press, 2000), 79.

25. Fuel Savings Society, *A History of the Fuel Savings Society*, 10; UBA, *Second Annual Report of the Executive Board of the Union Benevolent Association* (Philadelphia: J. Harding, 1833), 3; UBA, *The Seventh Annual Report of the Union Benevolent Association, Read at the Annual Meeting, October 23, 1838* (Philadelphia: J. Van Court, 1838), 4; minutes of 15 Nov. 1842 and 25 Nov. 1845 meetings of the executive board, in "Minutes of the Executive Board, 1831–1854," UBA Records, HSP; UBA, *Eighteenth Annual Report of the Executive Board, and of the Ladies' Board of Managers of the Union Benevolent Association* (Philadelphia, 1849), 5.

26. *Address to the Public by the Lackawaxen Coal Mine and Navigation Company, Relative to the Proposed Canal from the Hudson to the Head Waters of the Lackawaxen River* (New York, 1824), 7–8; Dorothy Hurlbut Sanderson, *The Delaware and Hudson Canalway: Carrying Coals to Rondout* (Ellenville, NY: Rondout Valley, 1965), 3–4.

27. *New York Commercial Advertiser*, 30 Jan. 1828; Delaware and Hudson Canal Company, *Annual Report of the Board of Managers of the Delaware and Hudson Canal Company to the Stockholders for the Year 1831* (New York, 1832), 4; A Stockholder of the Morris Canal, *Letter to John Wurtz, Esq. with Case and Opinion* (New York, [1831]), 11. See also *A Review by a Stockholder of the Morris Canal, of the "Views of a Stockholder, in Relation to the Delaware and Hudson Canal Company"* (Jersey City, NJ, 1831).

28. *New York Evening Post*, 17 Jan., 26 Jan., Feb. 11, 1831; *New York Commercial Advertiser*, 17 Jan., 18 Jan., 21 Jan., 10 Feb., 1831.

29. *Delaware and Hudson Canal Company 1831 Annual Report*, 5–6; *New York Commercial Advertiser*, 25 Jan., 10 Feb. 1831; *New York Evening Post*, 27 Jan. 1831; John Wurts to Charles Wurts, 5 Feb. 1831, Maurice Wurts to Charles Wurts, 10 Feb. 1831, Wurts Family Papers, HSP.

30. *New York Commercial Advertiser*, 3 Jan., 10 Feb. 1831; *New York Evening Post*, 10 Feb. 1831; Maurice Wurts to Charles Wurts, 12 Feb. 1831, Wurts Family Papers, HSP.

31. Maurice Wurts to Charles Wurts, 12 Feb. 1831, Wurts Family Papers, HSP; UBA, *First Annual Report*, 7; UBA, *Seventh Annual Report*, 4; UBA, *Eighteenth Annual Report of the Executive Board, and of the Ladies Board of Managers of the Union Benevolent Association* (Philadelphia, 1849), 5–6.

32. Christopher Jones, "The Carbon-Consuming Home: Residential Markets and Energy Transitions," *Enterprise & Society* 12 (2011): 801; *Niles' Register*, 15 Nov. 1845.

33. *Relfs Philadelphia Gazette, and Daily Advertiser*, 31 Aug. 1836; James C. Neal, *Charcoal Sketches; or, Scenes from the Metropolis* (Philadelphia: T. B. Peterson and

Brothers, 1865), 93–99; David E. E. Sloane, "The Comic Writers of Philadelphia: George Helmbold's 'The Tickler,' Joseph C. Neal's 'City Worthies,' and the Beginning of Modern Periodical Humor in America," *Victorian Periodicals Review* 28 (1995): 186–198.

34. Neal, *Charcoal Sketches*, 93.

CHAPTER THREE: How the Coal Trade Made Heat Cheap

1. John Rockwell to Alfred Rockwell, 4 Feb., 5 Feb., 8 May 1867, Rockwell Family Papers, CHS.

2. John Rockwell to Alfred Rockwell, 5 Feb., 21 Feb., 3 Aug., 7 Sept. 1867, Rockwell Family Papers, CHS; *Chicago Tribune*, 11 Aug. 1867; David Montgomery, *Beyond Equality: Labor and the Radical Republicans* (New York: Alfred A. Knopf, 1967), 435–438.

3. Bayard Tuckerman, ed., *The Diary of Philip Hone, 1828–1851*, vol. 1 (New York: Dodd, Mead, 1889), 390.

4. *Register of Pennsylvania*, 7 Mar. 1835; Katherine Harvey, *Best-Dressed Miners: Life and Labour in the Maryland Coal Region, 1835–1910* (Ithaca, NY: Cornell University Press, 1970); Andrew Roy, "The Mines and Mining Resources of Ohio," in Henry Howe, ed., *Historical Collections of Ohio*, vol. 1 (Columbus, OH: Henry Howe and Son, 1889), 112–113; Howard Eavenson, *The First Century and a Quarter of American Coal Industry* (Pittsburgh: privately printed, 1942), 270; *Hunt's Merchants' Magazine* 32 (Feb. 1855): 252. The idea of "developmental canals" in the American North is best explained in John Majewksi, *A House Divided: Economic Development in Pennsylvania and Virginia before the Civil War* (New York: Cambridge University Press, 2000), 37–58.

5. *Hunt's Merchants' Magazine* 30 (Feb. 1854): 248; *Hunt's Merchants' Magazine* 45 (July 1861): 135; Christopher Jones, "Energy Landscapes: Coal Canals, Oil Pipelines, and Electricity Transmission Wires in the Mid-Atlantic, 1820–1930," (PhD diss., University of Pennsylvania, 2009), 86–88.

6. Jones, "Energy Landscapes," 115.

7. A good, basic introduction to these methods of coal mining in the United States is found in Priscilla Long, *Where the Sun Never Shines: A History of America's Bloody Coal Industry* (New York: Paragon House, 1989), 23–36.

8. Anthony F. C. Wallace, *St. Clair: A Nineteenth-Century Coal Town's Experience with a Disaster-Prone Industry* (Ithaca, NY: Cornell University Press, 1988), 39–53; Edmund Ruffin, "Notes of a Three-Day Excursion into Goochland, Chesterfield, and Powhatan," *Farmer's Register* 5 (1837): 316.

9. Long, *Where the Sun Never Shines*, 36–38.

10. Historian Thomas Andrews describes the "workscape" for coal miners in the western coalfields of Colorado quite aptly as a "constellation of unruly and ever-unfolding relationships." Thomas Andrews, *Killing for Coal: America's Deadliest Labor War* (Cambridge, MA: Harvard University Press, 2008), 125.

11. Wallace, *St. Clair*, 15–18.

12. "Coal Region of the Schuylkill and Wyoming Valley," *Hunt's Merchants' Magazine* 14 (June 1846): 540; Herman Haupt, *The Coal Business on the Pennsylvania Railroad* (Philadelphia: T. K. and P. G. Collins, 1857), 33; J. W. Foster, *Report upon the Mineral Resources of the Illinois Central Railroad* (New York: George Scott Roe, 1856), 22.

13. *Poulson's American Daily Advertiser* (Philadelphia), 6 May 1835; *Hazard's Register of Pennsylvania*, 14 Dec. 1833; *Longworth's American Almanac, 1815* (New York: Directory Office, 1815); *Longworth's American Almanac, 1830* (New York: Directory Office, 1830); *Doggett's New-York City Directory for 1845 and 1846* (New York: John Doggett, Jr., 1846); Francis Doremus Receipt Book, NYHS.

14. George G. Foster, *New York by Gas-Light and Other Urban Sketches*, ed. Stuart Blumin (Berkeley: University of California Press, 1990), 128–130.

15. *Poulson's American Daily Advertiser*, 6 May, 29 May, 29 June 1835; *Philadelphia Saturday Courier*, 13 June 1835; *Register of Pennsylvania*, 4 July 1835; William A. Sullivan, "A Decade of Labor Strife," *Pennsylvania History* 17 (1950): 23–38; Bruce Laurie, *Working People of Philadelphia, 1800–1850* (Philadelphia: Temple University Press, 1980), 90–91, 99; John Dupuy to Charles Trego, 9 Apr. 1846, Charles Trego file, American Philosophical Society Archives, Philadelphia.

16. William Cronon, *Nature's Metropolis: Chicago and the Great West* (New York: W. W. Norton, 1991), 266; A. T. Andreas, *History of Chicago from the Earliest Period to the Present Time*, 3 vols. (Chicago: A. T. Andreas, 1884–1886), 1:575–576, 2:330, 673; Bessie Louise Pierce, *A History of Chicago*, vol. 2, *From Town to City, 1848–1871* (New York: Alfred A. Knopf, 1940), 115, 466; Little Rock Mining Co., *A Statement of the Operations of the Little Rock Mining Co. in the La Salle Coal Basin* (Chicago: Charles Scott, 1858), 9, 11; Eavenson, *American Coal Industry*, 535.

17. Eavenson, *American Coal Industry*, 420, 433–434; Department of Commerce and Labor, *Statistical Abstract of the United States for 1907* (Washington, DC: Government Printing Office, 1908), 567; R. G. Healey, *The Pennsylvania Anthracite Coal Industry, 1860–1902* (Scranton, PA: University of Scranton Press, 2007), 198–200.

18. Healey, *Pennsylvania Anthracite Coal Industry*, 22; *Statistical Abstract of the United States for 1907*, 567; Eavenson, *American Coal Industry*, 402; *Chicago Tribune*, 9 July 1863, 3 Nov. 1865.

19. Supply Clap Thwing to Judah Tours, 5 Nov. 1847, to James S. Cox, 22 Feb. 1849, to David Marcy, 14 Oct. 1857, to Charles Heckscher, 25 July 1859, to Captain William Frost, 19 July 1862, Supply Clap Thwing Letterbooks, MHS; Supply Clap Thwing Invoice for Coal, 8 Apr. 1862, ibid.

20. S. J. Packer, *Report of the Committee of the Senate of Pennsylvania upon the Subject of the Coal Trade* (Harrisburg, PA: Henry Welsh, 1834), 24; George Heberton Evans Jr., *Business Incorporations in the United States, 1800–1943* (New York: National Bureau of Economic Research, 1948), 11; L. Ray Gunn, *The Decline of Authority: Public Economic Policy and Political Development in New York, 1800–1860* (Ithaca, NY: Cornell University Press, 1988), 222–245; Sean Patrick Adams, *Old Dominion, Industrial*

Commonwealth: Coal, Politics, and Economy in Antebellum America (Baltimore: Johns Hopkins University Press, 2004), 170–178.

21. Evans, *Business Incorporations in the United States*, 19; John W. Eilert, "Illinois Business Incorporations, 1861–1869," *Business History Review* 37 (1963): 179; C. B. Conant, "Coal Fever," *Merchants' Magazine and Commercial Review* 52 (May 1865): 350; Adams, *Old Dominion, Industrial Commonwealth*, 196–202; Alexander K. McClure, *Old Time Notes of Pennsylvania: A Connected and Chronological Record of the Commercial, Industrial and Educational Advancement of Pennsylvania, and the Inner History of All Political Movements since the Adoption of the Constitution of 1838*, vol. 1 (Philadelphia: John C. Winston, 1905), 535.

22. *Miner's Journal* (Pottsville, PA), 19 Mar., 26 Mar. 1864; Calvin G. Beitel, *A Digest of Titles of Corporations Chartered by the Legislature of Pennsylvania between the Years 1700 and 1873 Inclusive* (Philadelphia: John Campbell and Son, 1874), 275; James Macfarlane, *The Coal-Regions of America: Their Topography, Geology, and Development*, 3rd ed. (New York: D. Appleton, 1875), 664–665; Adams, *Old Dominion, Industrial Commonwealth*, 196–202; Isaac Costa, comp., *Gopsill's Pennsylvania State Business Directory: Containing the Names and Addresses of Merchants, Manufacturers, Professional Men, and over 70,000 Farmers. With an Appendix Giving a Complete List of Banks, Insurance Companies, Railroads, Corporations, Newspapers, and Other Useful Information* (Jersey City, NJ: James Gopsill, 1865), 911–914.

23. J. Walter Coleman, *Labor Disturbances in Pennsylvania, 1850–1880* (Washington, DC: Catholic University of America Press, 1936), 40–49; Arnold Shankman, "Draft Riots in Civil War Pennsylvania," *Pennsylvania Magazine of History and Biography* 101 (1977): 190–204; Grace Palladino, *Another Civil War: Labor, Capital, and the State in the Anthracite Regions of Pennsylvania, 1840–68* (Urbana: University of Illinois Press, 1990), 140–162; Kevin Kenny, *Making Sense of the Molly Maguires* (New York: Oxford University Press, 1998), 87–102; Long, *Where the Sun Never Shines*, 98–99; William A. Russ Jr., "The Origin of the Ban on Special Legislation in the Constitution of 1873," *Pennsylvania History* 11 (1944): 260–275; J. P. Shalloo, *Private Police: With Special Reference to Pennsylvania* (Philadelphia: American Academy of Political and Social Science, 1933), 58–65.

24. Andrew Roy, *A History of the Coal Miners of the United States, from the Development of the Mines to the Close of the Anthracite Strike of 1902 . . .* (Columbus, OH: J. L. Trauger, [1907]), 62–67; Edward Wieck, *The American Miners' Association: A Record of the Origin of Coal Miners' Unions in the United States* (New York: Russell Sage Foundation, 1940), 97–98, 136–161, 219, 226; David Montgomery, *Beyond Equality: Labor and the Radical Republicans, 1862–1972* (New York: Alfred A. Knopf, 1967), 92; John H. M. Laslett, *Colliers Across the Sea: A Comparative Study of Class Formation in Scotland and the American Midwest, 1830–1924* (Urbana: University of Illinois Press, 2000), 98.

25. *Illinois Farmer* 8 (1863): 119–120; Wieck, *American Miners' Association*, 238; Montgomery, *Beyond Equality*, 98–99.

26. George W. I. Ball, comp., *General Railroad Laws of the State of Pennsylvania and Acts Relative to Corporations Affecting Railroad Companies* (Philadelphia: Allen, Lane, and Scott, 1875), 98–100, 101–102, 110, 118–119, 123–127, 128–130; *Summit Branch Railroad Co. and the Bear Valley Coal Company. Charter, Capital Stock, &c., &c* (Boston, 1863), 6.

27. *Boston Daily Advertiser*, 25 May 1863; *North American and United States Gazette* (Philadelphia), 15 Dec. 1863.

28. Wieck, *American Miners' Association*, 107–111, quotations on 274, 284.

29. S. H. Sweet, *Special Report on Coal: Showing Its Distribution, Classification and Cost Delivered over Different Routes to Various Points in the State of New York, and the Principal Cities on the Atlantic Coast* (New York: D. Van Nostrand, 1866), 3; *Appeal of the Boatmen of the Schuylkill Canal, to the Coal Consumers, Coal Producers, and the Coal Land-Owners* (Port Carbon, PA: n.p., 1864), 5–6; *Chicago Tribune*, 3 Nov. 1865.

30. *The Consumers' Mutual Coal Company* (Philadelphia: H. Evans, 1864), 3–4, 7; Harmony Mutual Coal Company, *The Harmony Mutual Coal Company* (New York: J. O. Seymour, 1866), 2; *Housekeeper's Coal Company, 1864–65* (n.p., n.d.), Library of Congress; *Philadelphia Intelligencer* 8 (Nov. 1864): 83; *Scientific American*, 20 May 1865; *American Railroad Journal*, 21 Oct. 1865.

31. Charles Barnard, "From Hod to Mine. In Seven Lifts," *American Homes* (July 1874): 482, 484, 485.

32. Ibid., 487; Charles Barnard, "From Hod to Mine. In Seven Lifts," *American Homes* (Aug. 1874): 548, 550, 551.

33. Charles Barnard, "From Hod to Mine. In Seven Lifts," *American Homes* (Sept. 1874): 611, 612, 615, 617.

CHAPTER FOUR: How the Industrial Hearth Defied Control

1. Priscilla Brewer, *From Fireplace to Cookstove: Technology and the Domestic Ideal in America* (Syracuse, NY: Syracuse University Press, 2000), 175–177; Alfred J. Pairpoint, *Rambles in America, Past and Present* (Boston: Alfred Mudge and Sons, 1891), 165.

2. Susan Strasser, *Never Done: A History of American Housework* (New York: Pantheon Books, 1982), 54–55; Jeanne Boydston, *Home and Work: Housework, Wages, and the Ideology of Labor in the Early Republic* (New York: Oxford University Press, 1990), 86, 106–107; Daniel quoted in Brewer, *From Fireplace to Cookstove*, 177; *Ladies' Own Magazine*, 1 May 1870; Ruth Schwartz Cowen, *More Work for Mother: The Ironies of Household Technology from the Open Hearth to the Microwave* (New York: Basic Books, 1983), 61.

3. Brewer, *From Fireplace to Cookstove*, 46–52; John Crowley, *The Invention of Comfort: Sensibilities and Design in Early Modern Britain and Early America* (Baltimore: Johns Hopkins University Press, 2001), 182–190; David Handlin, *The American Home: Architecture and Society, 1815–1915* (Boston: Little, Brown, 1979), 56–58; "Bad Effects of Breathing Impure Air," *Boston Medical and Surgical Journal* 25 (11 Aug.

1841): 7; *Scientific American*, 3 Jan. 1852; Morrill Wyman, *A Practical Treatise on Ventilation* (Boston: James Munroe, 1846), 191; Luther Bell, *The Practical Methods of Ventilating Buildings, Being the Annual Address before the Massachusetts Medical Society, May 31, 1848* (Boston: Damrell and Moore, 1848), 26.

4. *The Regenerator, Otherwise Practical Progressive Philosopher* 4 (Jan. 1851): 289–291; S. Webber, "Further Remarks on Ventilation and the Warming of Rooms," *American Journal of Science and Arts* 14 (Sept. 1852): 184, 188, 190; D. M. Dewey, *Heat and Ventilation: General Observations on the Atmosphere and Its Abuses, as Connected with the Common or Popular Mode of Heating Public and Private Buildings, Together with Practical Suggestions for the Best Mode of Warming and Ventilating* (Rochester, NY: Arcade Hall, 1852).

5. Benjamin Walbert III, "The Infancy of Central Heating in the United States: 1803–1845," *Bulletin of the Association for Preservation Technology* 3 (1971): 76–88; Robert Bruegmann, "Central Heating and Forced Ventilation: Origins and Effects on Architectural Design," *Journal of the Society of Architectural Historians* 37 (1978): 143–160; *Nicholas Mason's Warming and Ventilation Warehouse, No. 3 Washington Street, Boston, Mass.* (Boston: Shepard, 1854), 5, 8–9; Walter Bryant, *A Treatise on the Heating and Ventilating of Dwelling Houses, School Houses, Churches, and all Kinds of Public Buildings* (Boston: J. H. and F. F. Farwell, 1861), 15.

6. "On Warming and Ventilating Houses," *Appletons' Mechanics' Magazine and Engineers' Journal* 3 (Sept. 1853): 213; Bryant, *A Treatise on Heating and Ventilating*, 3–4; William B. Meyer, "Harvard and the Heating Revolution," *New England Quarterly* 77 (2004): 602–604; *Report of the Committee on the Expediency of Providing Better Tenements for the Poor* (Boston: Eastburn's, 1846), 25; Edgar Martin, *The Standard of Living in 1860* (Chicago: University of Chicago Press, 1942), 90–94; William B. Meyer, *Americans and Their Weather* (New York: Oxford University Press, 2000), 77.

7. James Parton, "Cincinnati," *Atlantic Monthly* 20 (Aug. 1867): 234; *Cleveland Herald*, 5 Mar. 1872; James Macaulay, *Across the Ferry: First Impressions of America and Its People*, 2nd ed. (London: Hodder and Stoughton, 1872), 240–241; *Boston Daily Advertiser*, 13 Dec. 1867; Parton and Glazier quotes, respectively, from Angela Gugliotta, "How, When, and for Whom Was Smoke a Problem in Pittsburgh," in *Devastation and Renewal: An Environmental History of Pittsburgh and Its Region*, ed. Joel A. Tarr (Pittsburgh: University of Pittsburgh Press, 2003), 112; and David Stradling, *Smokestacks and Progressives: Environmentalists, Engineers, and Air Quality in America, 1881–1951* (Baltimore: Johns Hopkins University Press, 1999), 38; Alexander Craib, *America and the Americans: A Narrative of a Tour in the United States and Canada, with Chapters on American Home Life* (London: A. Gardner, 1892), 117–118.

8. Gugliotta, "How, When, and for Whom Was Smoke a Problem in Pittsburgh," 113; Stradling, *Smokestacks and Progressives*, 46–48; *Cleveland Herald*, 18 Jan. 1866; *St. Louis Globe-Democrat*, 23 Apr. 1881; R. Dale Grinder, "The Battle for Clean Air: The Smoke Problem in Post–Civil War America," in *Pollution and Reform in American Cities, 1870–1930*, ed. Martin Melosi (Austin: University of Texas Press, 1980), 83–103.

9. Anthony Jackson, *A Place Called Home: A History of Low-Cost Housing in Manhattan* (Cambridge, MA: MIT Press, 1976), 24–25; Gwendolyn Wright, *Building the Dream: A Social History of Housing in America* (New York: Pantheon Books, 1981), 117–188.

10. Paul Boyer, *Urban Masses and Moral Order in America, 1820–1920* (Cambridge, MA: Harvard University Press, 1978), 124; An Act for the Regulation of Tenement and Lodging Houses in the Cities of New York and Brooklyn, New York State Laws, Ninetieth Session, 1867, Chapter 908, Section 15; Richard Plunz, *A History of Housing in New York City: Dwelling Type and Social Change in the American Metropolis* (New York: Columbia University Press, 1990), 24–29; Wright, *Building the Dream*, 122–123.

11. *Report of the Massachusetts Bureau of Statistics of Labor, 1870* (Boston: Wright and Potter, 1870), 173, 176, 179, 246, 272; Associated Charities of Boston, *Laws Applying to Tenements in the City of Boston* (Boston: Associated Charities of Boston, 1889), 9.

12. George Derby, *An Inquiry into the Influence of Anthracite Fires upon Health*, 2nd ed. (Boston: A. Williams, 1868), 11, 52–53, 59–60, 65–66; Frederick Binder, "Anthracite Enters the American Home," *Pennsylvania Magazine of History and Biography* 82 (1958): 97–98; Barbara Gutmann Rosenkrantz, *Public Health and the State: Changing Views in Massachusetts, 1842–1936* (Cambridge, MA: Harvard University Press, 1972), 52–67, quote from p. 67.

13. Catharine Beecher and Harriet Beecher Stowe, *The American Woman's Home: Principles of Domestic Science* (New York: J. B. Ford, 1869), 49, 58–61, 63, 83, 429, 431; Strasser, *Never Done*, 56–57.

14. Joseph Lyman and Laura Lyman, *The Philosophy of Housekeeping: A Scientific and Practical Manual* (Hartford, CT: S. M. Betts, 1869), 471–473.

15. *City of Boston, Ordered That the Chief of Police Be Directed to Notify All Owners or Occupants of Coal Holes* (Boston: n.p., 1855); City of Boston, *Rules and Regulations in Relation to Coal-Holes, Vaults, &c. under the Sidewalks* (Boston: n.p., 1863); *Saward's Coal Trade Journal* (New York), 10 Jan. 1877.

16. Supply Clap Thwing to C. A. Heckscher, 2 Apr. 1850, Supply Clap Thwing Letterbooks, MHS; John Kirk to J. H. Hailman & Co., 6 Mar. 1867, John Kirk Letterbook, CHS; George Bowen to John Omsbre (?), 12 Aug. 1869, and George Bowen to Linnickson & Company, 1 Sept. 1869, George Bowen & Company Records, HML; Charles Miesse, *Points on Coal and the Coal Business Containing an Explanation of How Coal Was Formed, Coal Veins, How They Were Deposited.* (Myerstown, PA: Feese and Uhrich, 1887), 184.

17. *Chicago Tribune*, 14 Jan., 16 Jan., 4 July 1868, 30 July 1871; Bessie Louise Pierce, *A History of Chicago*, vol. 2, *From Town to City, 1848–1871* (New York: Alfred A. Knopf, 1940), 181–183. A national survey of retail coal dealers a half-century later found that profit margins continued to be thin; on average, profits ranged from 0.342 to 0.542 cents per ton. See Roderick Stephens, "The Margins of Retail Coal Dealers," *Annals of the American Academy of Political and Social Science* 111 (1924): 169.

18. *Chicago Tribune*, 12 Dec. 1866; *Coal and Iron Record*, 30 Mar. 1872.

19. George Bowen to Blakiston & Graeff, 13 Sept. 1869, to Linnickson & Co., 22

Sept. 1869, to John Omsbre (?), 5 Oct. 1869, to Blakison & Graeff, 1 Dec., 23 Dec. 1869, George Bowen & Company Records, HML; *Saward's Coal Trade Journal* (New York), 8 Apr. 1874.

20. R. C. Taylor, *Statistics of Coal; Including Mineral Bituminous Substances Employed in Arts and Manufactures*, 2nd ed. (Philadelphia: J. W. Moore, 1855), 233n; *Gray's New England Real Estate Journal*, 15 Feb. 1869; *Saward's Coal Trade Journal*, 15 Dec. 1875, 10 Jan., 16 May 1877; *Chicago Tribune*, 15 Nov. 1874; *Scientific American*, 17 Feb. 1877.

21. On licensing in retail trades, see William Novak, *The People's Welfare: Law and Regulation in Nineteenth-Century America* (Chapel Hill: University of North Carolina Press, 1996), 90–95; *Proceedings and Debates of the Constitutional Convention of the State of New York, Held in 1867 and 1868, in the City of Albany*, vol. 2 (Albany: Weed, Parsons, 1868), 1369, 1370; "Warming Houses," *Hall's Journal of Health* 23 (1876): 530.

22. *Laws of the General Assembly of Pennsylvania Passed at the Session of 1871* (Harrisburg: B. Singerly, 1871), 1287–1290; *Journal of the Common Council of the City of Philadelphia*, vol. 2 (Philadelphia: King and Baird, 1874), 355–356; R. G. Healey, *The Pennsylvania Anthracite Coal Industry, 1860–1902: Economic Cycles, Business Decision-Making and Regional Dynamics* (Scranton, PA: University of Scranton Press, 2007), 227.

23. "Philadelphia Retail Coal Dealers Association Circular, 1 September 1880," "Confidential Circular of the Philadelphia Retail Coal Dealers Association, 15 October 1880," "Private and Confidential Circular of the Philadelphia Retail Coal Dealers' Association, 18 July 1881," "Confidential Circular of the Philadelphia Retail Coal Dealers Association, 21 June 1882," "Annual Report of the Board of Managers of the Philadelphia Retail Coal Dealers' Association, 20 June 1883;" Accounts and Scrapbook File, Donaghy and Sons Papers, HSP; Miesse, *Points on Coal*, 185.

24. Andrew Arnold, *From the Mine to the Fire: Railroads and Coal in the Gilded Age, 1870–1902* (New York: New York University Press, forthcoming); John Bowman, *Capitalist Collective Action: Competition, Cooperation, and Conflict in the Coal Industry* (New York: Cambridge University Press, 1989), 93–132.

25. Donald L. Miller and Richard E. Sharpless, *The Kingdom of Coal: Work, Enterprise, and Ethnic Communities in the Mine Fields* (Philadelphia: University of Pennsylvania Press, 1985), 55–58, 152–153; *Journal of a Tour of the Board of Trade of the Coal Dealers' Association of Boston and Vicinity* (Boston: Nathan Sawyer and Son, 1868), 11; David Brody, *In Labor's Cause: Main Themes on the History of the American Worker* (New York: Oxford University Press, 1993), 139.

26. Anthony F. C. Wallace, *St. Clair: A Nineteenth-Century Coal Town's Experience with a Disaster-Prone Industry* (Ithaca, NY: Cornell University Press, 1988), 121, 334; Marvin Wilson Schlegel, *Ruler of the Reading: The Life of Franklin Gowen, 1836–1889* (Harrisburg, PA: Archives Publishing, 1947), 7–8, 15, 45; C. K. Yearley, *Enterprise and Anthracite: Economics and Democracy in Schuylkill County, 1820–1875* (Baltimore: Johns Hopkins Press, 1961), 58–59; *The Coal Monopoly: The Coal Trade of Philadelphia in Reply to the President of the Philadelphia and Reading Railroad Company* (Philadelphia: A. T. Ziesing, 1873), 4.

27. *Hunt's Merchants' Magazine* 61 (Sept. 1869): 170–171; R. G. Healey, *Pennsylvania Anthracite Industry*, 249–273; *The Coal Monopoly*, 8, 16.

28. Wallace, *St. Clair*, 421–427, quote from p. 427; Miller and Sharpless, *The Kingdom of Coal*, 155–158; Priscilla Long, *Where the Sun Never Shines: A History of America's Bloody Coal Industry* (New York: Paragon House, 1989), 108–110.

29. Miller and Sharpless, *The Kingdom of Coal*, 160–170; Kevin Kenney, *Making Sense of the Molly Maguires* (New York: Oxford University Press, 1998), 213–214.

30. Eliot Jones, *The Anthracite Coal Combination in the United States, with Some Account of the Early Development of the Anthracite Industry* (Cambridge, MA: Harvard University Press, 1914), 40–58; Perry Blatz, *Democratic Miners: Work and Labor Relations in the Anthracite Coal Industry, 1875–1925* (Albany: State University of New York Press, 1994), 38–40.

31. Schlegel, *Ruler of the Reading*, 286–287.

32. *Boston Daily Advertiser*, 29 May 1869; Charles Miesse, *Points on Coal*, 160, 161; Terence V. Powderly, *The Path I Trod: The Autobiography of Terence V. Powderly*, ed. Harry J. Carman, Henry David, and Paul N. Guthrie (New York: Columbia University Press, 1940), 190–192.

CHAPTER FIVE: How Steam Heat Found Its Limits

1. James Jones Walworth Diary, entries from 1 May 1829–24 July 1830, in J. J. Walworth Papers, MHS.

2. Walworth Diary, entries from 8 Jan. 1846–29 Mar. 1832, MHS; Walworth Company, *Walworth, 1842–1942* (New York: Walworth, ca. 1945), 14–15, 22, 60; Ara Marcus Daniels, "A History of the Development of the Radiator," *Domestic Engineering* 130 (1930): 54–56. The apocryphal Andrew Jackson story appears in the trade press for steam heating. See, for example, the Walworth Company's *Walworth, 1842–1942*, 60.

3. Benjamin Walbert III, "The Infancy of Central Heating in the United States: 1803–1845," *Bulletin of the Association for Preservation Technology* 3 (1971): 78–80; Eugene S. Ferguson, "An Historical Sketch of Central Heating, 1800–1860," in *Building Early America: Contributions toward the History of a Great Industry*, ed. Charles E. Peterson (Radnor, PA: Chilton, 1976), 165–185; *New York Municipal Gazette*, 31 Dec. 1847.

4. Charles Davenport, *Descriptive Notice of the Steam Heating Apparatus for Stores, Dwelling Houses, and Public Buildings* (Boston: Journal Press, 1855), 13, 15; *Descriptive Notice of Wm. G. Pike & Co's Low Pressure Steam Heating Apparatus, for Stores, Dwelling Houses and Public Buildings, Patented June 23, 1857* (Boston: Bazin and Chandler, 1857), 13; Massachusetts Steam Heating Company, *Gold's Patent Steam Heating Apparatus for Warming Private Residences, Stores, Churches, Hospitals, Public Buildings, Green Houses, Graperies, &c* (Boston: Massachusetts Steam Heating Company, 1858), 9–10, 15, 18; New-York Steam Heating Co., *Proprietors of Gold's Patent Low Pressure Self-Regulating Steam Heating Apparatus* (New York: W. H. Arthur, 1859), 21; Susan

Reed Stifler, *The Beginnings of a Century of Steam and Water Heating by the H. B. Smith Company* (Westfield, MA: H. B. Smith, 1960), 26–29.

5. *Appletons' Mechanics' Magazine and Engineers' Journal* 3 (1853): 189; *Descriptive Notice of Wm. G. Pike & Co*, 13; *Waverly Magazine and Literary Repository*, 17 Oct. 1857.

6. *The Union Warming and Ventilating Company* (Boston: Geo. S. Rand, 1865), 5; Baker, Smith & Co., *Household Comfort: Artificial Warmth and Ventilation* (New York: Baker, Smith & Co., 1865), 3, 31, 38, 45, 64; Gold Heating Company, *How Shall We Heat Our Houses?* (New York: E. G. Groesbeck, 1872), 2.

7. Chauncey Depew, ed., *One Hundred Years of American Commerce*, vol. 2 (New York: D. O. Hayes, 1895), 361, 368; Stifler, *The Beginnings of a Century of Steam and Water Heating*, 72–73; Walworth Diary, entries for 31 May, 2 June, 8 June 1859, MHS.

8. Arthur Ducat, *The Practice of Fire Underwriting, Written and Compiled by Arthur C. Ducat, Late Secretary and Chief Surveyor of the Chicago Board of Underwriters, and Inspector General of the Army of the Cumberland* (New York: T. Jones, 1866), 56–57; *Boston Journal of Chemistry*, 1 Mar. 1873; *Engineering and Mining Journal* (New York), 21 Jan. 1873; "The Steam Pipe Incendiary," *Spectator* 10 (Feb. 1873): 96; F. C. Moore, *Fires: Their Causes, Prevention, and Extinction, Combining Also a Guide to Agents Respecting Insurance Against Loss by Fire and Containing Information as to the Construction of Buildings, Special Features of Manufacturing Hazards, Writing of Policies, Adjustment of Losses, Etc. Etc.* (New York: Continental Insurance Company of New York, 1877), 250–252.

9. *Scientific American* 19 (29 July 1868): 73; "The Fireplace Again," *Harvard Advocate* 7 (26 Mar. 1869), 42; "Pneumonia and Steam-Coils," *Hall's Journal of Health* 22 (Mar. 1875): 89; William B. Meyer, "Harvard and the Heating Revolution," *New England Quarterly* 77 (2004): 604–606. Silliman's comments come from a pamphlet appended to *The New York Steam Heating Company, Proprietors of Gold's Patent Low Pressure Self-Regulating Steam Heating Apparatus* (New York: W. H. Arthur, 1859), 47. This essay was a common feature of prospectuses and trade circulars for Gold's patented steam heating system.

10. Ruth Schwartz Cowan, *More Work for Mother: The Ironies of Household Technology from the Open Hearth to the Microwave* (New York: Basic Books, 1983), 96; Gwendolyn Wright, *Building the Dream: A Social History of Housing in America* (New York: Pantheon Books, 1981), 138–139. See also Susan Strasser, *Never Done: A History of American Housework* (New York: Pantheon Books, 1982), 53–56; Thomas Schlereth, "Conduits and Conduct: Home Utilities in Victorian America, 1876–1915," in *American Home Life, 1880–1930* ed. Jessica H. Foy and Thomas J. Schlereth (Knoxville: University of Tennessee Press, 1992), 227–229.

11. For an overview of the idea of the "networked city," see the essays in Joel A. Tarr and Gabriel Dupuy, eds., *Technology and the Rise of the Networked City in Europe and America* (Philadelphia: Temple University Press, 1988). Christine Rosen discusses the political problems inherent in reworking the urban landscape in Boston, Baltimore, and Chicago—even in the wake of devastating fires. See Christine Rosen, *The Limits of Power: Great Fires and the Process of City Growth in America* (New York: Cambridge University Press, 1986).

12. The best articulation of this view of the railroads bringing rationality and order to American capitalism is Alfred D. Chandler Jr., *The Visible Hand: The Managerial Revolution in American Business* (Cambridge, MA: Belknap, 1977), 120. Although still influential, Chandler's ideas about the rationalization of American business have undergone significant revision. The literature is huge, but for a brief assessment of how Chandler's model has been revised, see Richard John, "Elaborations, Revisions, Dissents: Alfred D. Chandler, Jr.'s, *The Visible Hand* after Twenty Years," *Business History Review* 71 (1997): 151–200.

13. Leslie Tomory, *Progressive Enlightenment: The Origins of the Gaslight Industry, 1780–1820* (Cambridge, MA: MIT Press, 2012), 5; Thomas Hughes, *Networks of Power: Electrification in Western Society, 1880–1930* (Baltimore: Johns Hopkins University Press, 1983), 465.

14. Richard John, *Network Nation: Inventing American Telecommunications* (Cambridge, MA: Belknap, 2010), 7; David E. Nye, *Consuming Power: A Social History of American Energies* (Cambridge, MA: MIT Press, 1998), 96.

15. Arthur Brisbane, *Social Destiny of Man; or, Association and Reorganization of Industry* (Philadelphia: C. F. Stollmeyer, 1840), 77; Herman Haupt, *Report of General Herman Haupt, C.E., on the Holly District of Steam Heating for Cities and Villages* (Lockport, NY: n.p., 1879), 11–17, 23, 78–79.

16. James Herbert Bartlett, *District Steam Supply: Heating Buildings by Steam from a Central Source* (Montreal: John Lovell and Son, 1884), 15; Arthur H. Cole, "Underground Social Capital," *Business History Review* 42 (1968): 487–489. A thorough history of district heating can be found in Morris A. Pierce, "The Introduction of Direct Pressure Water Supply, Cogeneration, and District Heating in Urban and Institutional Communities, 1863–1882," unpublished 1993 manuscript in HML.

17. New York Steam Corporation, *Fifty Years of New York Steam Service: The Story of the Founding and Development of a Public Utility* (New York: New York Steam Corporation, 1932), 9–29; *Sanitary Engineer* (New York), 25 Jan. 1883; Thomas Hughes, *Networks of Power: Electrification in Western Society, 1880–1930* (Baltimore: Johns Hopkins University Press, 1983), 40–45.

18. *Scientific American*, 19 Nov. 1881; *American Architect and Building News*, 16 Dec. 1882; *Puck*, 27 Dec. 1882.

19. *American Architect and Building News*, 2 Dec. 1882, 10 Mar. 1883; *Scientific American*, 22 Dec. 1888.

20. New York Steam Corporation, *Fifty Years of New York Steam Service*, 41–43; Haupt, *Report of General Herman Haupt*, 10.

21. *New York Times*, 14 Mar., 15 Mar., 18 Mar. 1888; New York Steam Corporation, *Fifty Years of New York Steam Service*, 43–44.

22. *New York Times*, 10 July, 16 July, 18 July 1890; *Christian Union* (New York City), 17 July 1890; *American Architect and Building News*, 19 July 1890.

23. Boston Heating Company, *Heat and Power Supplied from a Central Station: Advantage of the System* (Boston: n.p., 1889), 4; W. H. Pearce, "The Transmission of Steam Heat from Central Plants," *Heating and Ventilating Magazine* 14 (July 1917):

26–27; Albert Bigelow Paine, *In One Man's Life: Being Chapters from the Personal and Business Career of Theodore N. Vail* (New York: Harper and Brothers, 1921), 190–194.

24. American District Steam Company, *Holly System of Heating for Cities and Villages through Pipes Laid in the Public Streets* (Lockport, NY: American District Steam Company, 1890), 4, 9; T. M. Zeiders, "History of the Harrisburg Steam Heat & Power Company" (1922?), 1–2, from the records of the Harrisburg Steam Heat & Power Company, box 16, folder 16, in the Pennsylvania Power & Light Company Records (accession 1552), HML. Collection hereafter referred to as PP&LCR.

25. Zeiders, "History of the Harrisburg Steam Heat & Power Company," 4; "Presidents' Report of 18 June 1888," 51, Minute Book Volume for the Harrisburg Steam Heat & Power Company, PP&LCR.

26. "President's Report, 17 June 1889," 80, Minute Book Volume for the Harrisburg Steam Heat & Power Company, PP&LCR.

27. Ibid.; "Special Meeting of the Board of Directors, 14 September 1889," 85, and "Directors' Meeting of 14 August 1890," 102, Minute Book Volume for the Harrisburg Steam Heat & Power Company, PP&LCR.

28. Zeiders, "History of the Harrisburg Steam Heat & Power Company," 4–10.

29. Emmanuelle Gallo, "Skyscrapers and District Heating, an Inter-related History, 1876–1933," *Construction History* 19 (2003): 87–106; Cole, "Underground Social Capital," 487.

30. *Scientific American*, 19 Nov. 1881.

31. Mark H. Rose, *Cities of Light and Heat: Domesticating Gas and Electricity in Urban America* (University Park: Pennsylvania State University Press, 1995), 111–146.

32. Charles A. Thrall, "The Conservative Use of Modern Household Technology," *Technology and Culture* 23 (1982): 194.

33. William T. Stead, *If Christ Came to Chicago: A Plea for the Union of All Who Love in the Service of All Who Suffer* (Chicago: Laird and Lee, 1894), 424.

Epilogue

1. Automatic Heating and Lighting Company, *Something New: Comforts and Conveniences for Every Family* (Philadelphia: Automatic Heating and Lighting, [1873]), 22–23.

2. A. R. Heath, *Florence Oil Stoves for Summer and Winter Cooking, Heating and Illuminating, Manufactured by the Florence Sewing Machine Co.* (Hartford, CT: A. R. Heath, 1876), 3; Ellen Richards, *The Cost of Living as Modified by Sanitary Science*, 2nd ed. (New York: John Wiley and Sons, 1903), 63.

3. Joel Barton, "A Picturesque Outlaw," *Frank Leslie's Popular Monthly* 27 (June 1889): 676, 677, 679; Joseph Pickering Putnam, *The Open Fireplace in All Ages* (Boston: Ticknor, 1886), 87; Clare Brunce, "Aestheticism in Heat," *Harper's Bazaar*, 2 Mar. 1895. Architect David Handlin argues that Putnam's view of the fireplace tapped into a romantic view of older homes common during the era. David P. Handlin, *The American Home: Architecture and Society, 1815–1915* (Boston: Little, Brown, 1979), 481–486.

4. Marcus Reynolds, *The Housing of the Poor in American Cities: The Prize Essay of the American Economic Association for 1892* (Baltimore: American Economic Association, 1893), 247–248; *The Tenement House Act* (New York: Tenement House Department, 1903), 45; Robert Coit Chapin, *The Standard of Living among Workingmen's Families in New York City* (New York: Russell Sage Foundation, 1909), 115–116; *First Report of the Tenement House Department of the City of New York*, vol. 1 (New York: Martin B. Brown, 1903), 8.

5. Richard Nixon, "The Energy Emergency," reprinted in Karen R. Merrill, *The Oil Crisis of 1973–1974: A Brief History with Documents* (Boston: Bedford / St. Martin's, 2007), 66–71. Quotes are from pp. 66, 68.

6. *New York Times*, 5 Oct. 1902; Roosevelt quoted in Robert Cornell, *The Anthracite Coal Strike of 1902* (Washington, DC: Catholic University of America Press, 1957), 178.

SELECTED FURTHER READING

Most historians who examine the process of industrialization in American society use specific case studies. There are a few good surveys, however, that tackle the issue broadly. One excellent place to start is Walter Licht's *Industrializing America: The Nineteenth Century* (Baltimore: Johns Hopkins University Press, 1995). From there, readers can follow the themes discussed in this book through several excellent books and articles. While the short essay that follows is not comprehensive, it provides at least a good start.

Readers looking for more detail on the development of fireplaces and stoves have plenty of sources to consult. The eminent scholar Arthur Cole noted the scant understanding that historians have of the organic energy network in his article "The Mystery of Fuel Wood Marketing in the United States," *Business History Review* 44 (1970): 339–359. Two rather traditional narrative accounts of home heating chronicle the various improvements over the years that provide sweeping accounts of the subject with an emphasis upon the variety of designs and approaches are Josephine Peirce, *Fire on the Hearth: The Evolution and Romance of the Heating-Stove* (Springfield, MA: Pond-Ekberg, 1951), and Lawrence Wright, *Home Fires Burning: The History of Domestic Heating and Cooking* (London: Routledge, 1964). Priscilla Brewer offers a thoughtful approach to the subject in her work on the cookstove, *From Fireplace to Cookstove: Technology and the Domestic Ideal in America* (Syracuse, NY: Syracuse University Press, 2000), which discusses heating devices in their social and cultural context. On the business side, Howell Harris examines the growth and stagnation of the stove trade over the nineteenth century in his article "Inventing the US Stove Industry, c.1815–1875: Making and Selling the First Universal Consumer Durable," *Business History Review* 82 (2008): 701–733.

There is a growing literature on historical energy transitions in the United States. In dealing with the move from organic to mineral fuel, one school of historians highlights the contributions of specific individuals or institutions in spurring change. Three examples of this entrepreneurial take on the rise of coal include Clifton K. Yearley, *Enterprise and Anthracite: Economics and Democracy in Schuylkill County, 1820–1875* (Baltimore: Johns Hopkins Press, 1961); Alfred Chandler, "Anthracite Coal and the Beginnings of the Industrial Revolution in the United States," *Business History Review* 46 (1972): 141–181; and H. Benjamin Powell, *Philadelphia's First Fuel Crisis: Jacob Cist and the Developing Market for Pennsylvania Anthracite* (University Park: Pennsylvania State University Press, 1978). The rise of coal is sometimes depicted as a "revolution" in fuel use that transformed society from using renewable resources to using nonrenewable ones, from using living fuels to using fossil fuels, or from using relatively clean-burning fuels to using absolutely filthy ones. See, for example, Sam H. Schurr and Bruce C. Netschert, *Energy in the American Economy, 1850–1975: An Economic Study of Its History and Prospects* (Baltimore: Johns Hopkins Press, 1960); Frederick Binder, *Coal Age Empire: Pennsylvania Coal and Its Utilization to 1860* (Harrisburg: Pennsylvania Historical and Museum Commission, 1974); Barbara Freese, *Coal: A Human History* (Cambridge, MA: Perseus, 2003); and Alfred Crosby, *Children of the Sun: A History of Humanity's Unappeasable Appetite for Energy* (New York: W. W. Norton, 2006). Another approach, building upon a concept outlined by environmental historian Martin Melosi in 1982, views energy transition as an "evolutionary" rather than "revolutionary" process. See Martin Melosi, "Energy Transitions in the Nineteenth-Century Economy," in *Energy and Transport: Historical Perspectives on Policy Issues*, ed. George H. Daniels and Mark H. Rose (Beverley Hills, CA: Sage, 1982), 55–69; Vaclav Smil, *Energy in World History* (Boulder, CO: Westview, 1994); and David Nye, *Consuming Power: A Social History of American Energies* (Cambridge, MA: MIT Press, 1998). Two very long-term perspectives are in Rolf Peter Sieferle, *The Subterranean Forest: Energy Systems and the Industrial Revolution* (London: White Horse, 2010), and Roger Fouquet, *Heat, Power and Light: Revolutions in Energy Services* (Northampton, MA: Edward Elgar, 2010).

There are plenty of studies of the American coal industry, but most of them tend to focus on the mining regions and the conflict between miners and managers without delving into the wider connections. Good overviews include Donald L. Miller and Richard E. Sharpless, *The Kingdom of Coal: Work, Enterprise, and Ethnic Communities in the Mine Fields* (Philadelphia: University of Pennsylvania Press, 1985), and Priscilla Long, *Where the Sun Never Shines: A History of America's Bloody Coal Industry* (New York: Paragon House, 1989). The work of Thomas Andrews is an exception to this trend in coal mining history, offering a multifaceted examination of the growth of the Colorado coal industry in *Killing for Coal: America's Deadliest Labor War* (Cambridge, MA: Harvard University Press, 2008). The Colorado coal fields were not really a part of the story of *Home Fires*, but Andrews's characterization of the organic-to-mineral energy transition makes some excellent points that readers will find invaluable for understanding the topic. On the anthracite trade and the recurring curse of abundance, see Anthony F. C. Wallace's sociological study of an anthracite coal mining community, *St. Clair: A Nineteenth-Century Coal Town's Experience with a Disaster-Prone Industry* (Ithaca, NY: Cornell University Press, 1981).

The burning of coal in the American home raises important questions of work, gender, and the very loaded concept of "domesticity" in the nineteenth century. Jeanne Boydston, *Home and Work: Housework, Wages, and the Ideology of Labor in the Early Republic* (New York: Oxford University Press, 1990); Susan Strasser, *Never Done: A History of American Housework* (New York: Pantheon, 1982); and Ruth Schwartz Cowen, *More Work for Mother: The Ironies of Household Technology from the Open Hearth to the Microwave* (New York: Basic Books, 1983), each deal with different time periods in American history but share the same interest in exploring how women negotiated the home as workers, with technological and organizational challenges that were not exactly the same but were no less jarring than those faced by their male partners within the household. The significance of domestic markets to the larger issue of energy use is the subject of Christopher Jones's excellent article "The Carbon-Consuming Home: Residential Markets and Energy Transitions," *Enterprise & Society* 12 (Dec. 2011): 790–823. Once it was burned to generate warmth, coal still created problems for American cities, most notably in the form of pollution. For an overview of the movement to combat these problems, see David Stradling, *Smokestacks and Progressives: Environmentalists, Engineers, and Air Quality in America, 1881–1951* (Baltimore: Johns Hopkins University Press, 1999).

Finally, there are several important works that address the issue of "networked technology." See, for example, Leslie Tomory, *Progressive Enlightenment: The Origins of the Gaslight Industry, 1780–1820* (Cambridge, MA: MIT Press, 2012), on one of the earliest examples of this phenomenon at work. Another example is found in the story of the rise of electric power systems during the late nineteenth century as portrayed in Thomas Hughes's landmark study, *Networks of Power: Electrification in Western Society, 1880–1930* (Baltimore: Johns Hopkins University Press, 1983). This work focuses on the interaction of a series of "inventor-entrepreneurs," with Thomas Edison as the most famous example, and the economic and political contexts in the United States, Germany, and Great Britain. Above all, Hughes views the rise of electrical power in cities as an example of the emergence of distinct "sociotechnological" systems that follow a distinct course in which they are invented, overcome technical obstacles, grow, and eventually gain momentum. Finally, for an excellent of account of how networked technologies operated to deliver home heating in the period following *Home Fires*, see Mark Rose, *Cities of Light and Heat: Domesticating Gas and Electricity in Urban America* (University Park: Pennsylvania State University Press, 1995).

INDEX